Economic Analysis of Diversity in Modern Wheat

Economic Analysis of Diversity in Modern Wheat

Editors

Erika C.H. Meng

Economist

Impacts Targeting and Assessment Program

International Maize and Wheat Improvement Center (CIMMYT)

El Batán, Texcoco

Mexico

John P. Brennan

Principal Research Scientist

E.H. Graham Center for Agricultural Innovation

NSW Department of Primary Industries

Wagga Wagga, New South Wales

Australia

Science Publishers
www.scipub.net
234 May Street
Post Office Box 699
Enfield, New Hampshire 03748
United States of America

General enquiries : *info@scipub.net*
Editorial enquiries : *editor@scipub.net*
Sales enquiries : *sales@scipub.net*

Published by Science Publishers, Enfield, NH, USA
An imprint of Edenbridge Ltd., British Channel Islands
Printed in India

ISBN: 978-1-57808-575-0

Library of Congress Cataloging-in-Publication Data

Economic analysis of diversity in modern wheat/editors, Erika C.H. Meng, John P. Brennan.
p. cm.
Includes bibliographical references and index.
ISBN 978-1-57808-575-0 (hardcover)
1. Wheat-Economic aspects-China. 2. Wheat-Economic aspects-Australia. 3. Wheat-Varieties-China. 4. Wheat-Varieties-Australia. 5. Wheat-China-Genetics. 6. Wheat-Australia-Genetics. 7. Agriculture and state-China. 8. Agriculture and state-Australia. I. Meng, Erika C.H. II. Brennan, John P.
HD9049.W5C5936 2009
338.1'73110951-dc22

2008041305

Dedication

This book is dedicated to Erika Ching-Huei Meng (1963-2008), whose intelligence contributed so much to our understanding of the value of plant genetic resources, and whose unmatched courage will endure in our memories. Economist, colleague, athlete, linguist, and our friend, Erika tackled insurmountable challenges with integrity and grace.

Staff at the Center for Chinese Agriculture Policy also express their admiration for Erika's corage and integrity. Her passion for her career and her concern for and contributions to China's agricultural development has left deep impressions.

Royalties from this book will be distributed to the Erika C.H. Meng Scholarship at the University of California, Davis, where Erika completed her PhD. The fund supports graduate students in bridging applied development research and policy, as Erika did so well in her life.

Acknowledgements

The research presented in this book was conducted as part of the project "The Impact of Genetic Diversity on Wheat Crop Productivity: A Comparative Analysis of China and Australia." The editors and authors gratefully acknowledge the financial support of the Australian Centre for International Agricultural Research (ACIAR). Institutional support was also provided by the International Maize and Wheat Improvement Center (CIMMYT), the New South Wales Department of Primary Industries, and the Center for Chinese Agricultural Policy (CCAP), Chinese Academy of Sciences. The editors would like to express gratitude to Prabhu Pingali and John Dixon for their support of the research and to Melinda Smale for her helpful suggestions on the draft and her broader contributions in advancing the intellectual framework for the research and ideas for its application to China and Australia. We are also grateful to Paul Heisey for very useful comments and to Jorge Franco for his assistance with the analysis of morphological characteristics in both Australia and China. Kelly Cassaday and Mike Listman at CIMMYT provided valuable editorial advice and assistance. Finally, we would like to acknowledge the contributions and enthusiasm of the entire project research team, Melinda Smale, Ruifa Hu, Jikun Huang, Scott Rozelle, Songqing Jin, Ping Qin, David Godden, and Adam Bialowas. We are grateful to have had the opportunity to collaborate with all of you.

Foreword

Economic Analysis of Diversity in Modern Wheat

The term "genetic diversity" with reference to food crops was coined in the 1960s, but humans have mindfully molded natural crop diversity since the advent of agriculture. Hexaploid wheat, used to make popular foods like bread and Chinese noodles, appears to have reached humanity through repeated chance crosses of tetraploid wheat with wild grasses thousands of years ago. Since then intense farmer selection has domesticated and refined both forms of wheat, making the crop a key source of carbohydrates and other nutrients for millions worldwide. Science-based breeding in the twentieth century greatly accelerated wheat's evolution, producing high-yielding varieties that helped avoid famine in many developing countries. Emerging scientific tools hold promise for identifying and tapping new, useful genetic diversity within wheat's primary and secondary gene pools and, through genetic engineering, beyond.

But the rapid replacement of many wheat landraces with relatively few improved varieties on large expanses has raised concerns that this narrows genetic diversity. Recent experience—the emergence and spread of a new, virulent strain of stem rust from eastern Africa—seems to underline the need for broad diversity, both within and among varieties, as a frontline defense against evolving pathogens. Wheat genetic diversity has also been cited as a potential source of traits like heat tolerance, which can help wheat adapt to changing climates.

For these and other reasons, few would question diversity's value in general terms. The real issue is how much society at large is willing to pay for it. After all, the central questions confronting agriculture and finance policy makers dealing with food crops, especially in developing countries, are economic: What is diversity worth? Must the conservation of crop genetic diversity—a medium-to-long-term prospect—always come at the cost of reduced crop productivity in the near term?

To address those and related issues, the editors of this book have elected to focus on the circumstances and uses for wheat in Australia and China. The rich contrasts and intriguing analogies surrounding wheat in those settings furnish a useful lens for such an analysis. The book describes generally how policy affects wheat genetic diversity; it looks at historical changes in wheat genetic diversity, as policy and priorities have evolved; it identifies factors that explain changes and differences in spatial diversity; and finally, it analyzes the productivity impacts of changes in diversity. As a basis for discussion, the opening chapters define various types of crop genetic diversity and ways to measure them, framing the definitions and metrics in the contexts for which they are most relevant.

Wheat genetic diversity has been the topic of numerous studies by the Australian Centre for International Agricultural Research (ACIAR), the International Maize and Wheat Improvement Center (CIMMYT), and organizations like the International Food Policy Research Institute (IFPRI). Authors of the various chapters in this book have participated in those studies, or come from other advanced research institutes of recognized authority or national research organizations that are particularly well placed to address the theme. The present work reflects their pooled knowledge and concerns, and should prove of interest to a diverse audience including crop breeders, agricultural socioeconomists, research directors, and policy makers. We hope you find it interesting and useful, and welcome any comments you might have.

Masa Iwanaga
Director General
CIMMYT, 2002-08

Contents

Project Team/List of Contributors

Erika C.H. Meng
Economist
Impacts Targeting and Assessment Program
International Maize and Wheat Improvement Center
El Batán, Texcoco, Mexico

John P. Brennan
Principal Research Scientist
E.H. Graham Center for Agricultural Innovation
NSW Department of Primary Industries
Wagga Wagga, New South Wales, Australia

Melinda Smale
Senior Research Fellow
Environment and Production Technology
International Food Policy Research Institute
Washington, D.C., USA

Jikun Huang
Director
Center for Chinese Agricultural Policy
Institute of Geographical Sciences and Natural Resources Research
Chinese Academy of Sciences
Beijing, China

Ruifa Hu
Professor
Center for Chinese Agricultural Policy
Institute of Geographical Sciences and Natural Resources Research
Chinese Academy of Sciences
Beijing, China

Scott Rozelle
Helen F. Farnsworth Senior Fellow
Freeman Spogli Institute for International Studies
Stanford University
Palo Alto, California, USA

David Godden
Honorary Associate
Agricultural and Resource Economics
Faculty of Agriculture, Food & Natural Resources
University of Sydney
Sydney, New South Wales, Australia

Adam Bialowas
Centre for Microeconomics
University of Melbourne
Parkville, Victoria, Australia

Songqing Jin
Economist
Rural Development Research Group
World Bank
Washington, D.C., USA

Qin Ping
Graduate Student
Center for Chinese Agricultural Policy
Institute of Geographical Sciences and Natural Resources Research
Chinese Academy of Sciences
Beijing, China

Chapter 1

Role of Economics in Crop Genetic Diversity

E. Meng, J.P. Brennan, and M. Smale

ABSTRACT

Agricultural biodiversity is a key issue confronting farmers, scientists and policy makers. Economic analysis has much to contribute to the questions of value and behavior that are at the heart of the issues surrounding agricultural biodiversity. To ensure that appropriate policies are put in place to produce the desirable outcomes, a framework encompassing the multiple aspects of agricultural biodiversity and its measurement and impacts is needed. Analysis of different wheat production systems in China and Australia provides insights into the analytical framework and the implications at various levels of analysis of changes in crop diversity, an important component of agricultural biodiversity.

Scientific, public, and governmental awareness of the role of agricultural biodiversity and the need to ensure availability of and future access to genetic resources has increased considerably in recent decades. This attention has occurred amidst growing concerns that population pressure, urbanization and other changes in land use, and replacement of local varieties with improved varieties have resulted in the erosion of diversity and potentially risky levels of genetic uniformity. Concern about the conservation of agricultural biodiversity was initially motivated in part by fear that potentially valuable genes or genetic combinations would disappear from farmers' fields as higher-yielding modern varieties were adopted (Harlan 1972; Frankel 1970). The loss of farmer knowledge has also been raised as an irreversible consequence. However, as always, trade-offs exist, and one particularly significant issue in this debate is the extent to which society, or certain individuals in society, should forego welfare benefits from today's productivity gains for the uncertain benefit to future generations of producers and consumers.

Crop biodiversity plays an important role in both current production levels and future production possibilities (Cassman et al. 2005). Past negative experiences with crop uniformity have fostered the desire to avoid similar problems in the future, as well as the recognition that genetic resources are required for future use in improving yields and overcoming often unforeseen production constraints. Insufficient levels of crop diversity can potentially compromise the ability of natural systems and of scientists and farmers to respond to new pests, pathogens, and adverse environmental conditions. Crop scientists have long been concerned that widespread cultivation of genetically uniform varieties will increase vulnerability of the crop to biotic and abiotic stresses, so that the cost of uniformity can potentially be very high. A commonly cited example from the past is the southern corn leaf blight that resulted in serious economic damage in the United States in 1970. However, the risks are not limited to the past. An ex ante assessment of the potential risk of a new virulent race of wheat stem rust (U99) estimates that large amounts of the current wheat area in the potential risk zone from eastern Africa to eastern India are susceptible and that additional areas are also highly likely to be susceptible, largely due to high levels of uniformity (Singh et al. 2006).

Unforeseen challenges to long-term food security may also become increasingly difficult to address should the risk to particular genes and gene complexes rise. In the case of U99, the major sources of stem rust resistance in existence have been overcome after more than 30 years of effective protection for a large proportion of the world's wheat cultivars (Singh et al. 2006). An increasingly large body of research highlights the contributions of past utilization of genetic resources in crop improvement to achieving yield gains, maintaining yields through better disease resistance, and increasing crop adaptability to heterogeneous environments (Day Rubenstein et al. 2005; Fowler et al. 2001; Jana 1999; Hoisington et al. 1999; Cox et al. 1988).

A key unanswered question is whether trade-offs between diversity and immediate production objectives are inevitable. For the governments of some developing countries, the choice between the needs of today's consumers and producers and the uncertain welfare benefits to future generations may be unequivocal. Due to concerns for food security, increasing and sustaining current yield levels over time regularly top the list of agricultural research priorities. More than 80% of wheat area in the developing world is currently planted with scientifically improved wheat varieties released by plant breeding programs (Lantican et al. 2005; Heisey et al. 2002). Nevertheless, an important issue to address is whether the governments of those countries should also concern themselves with the conservation and management of crop genetic diversity, even if it conflicts with more immediate production objectives. To the extent that future production may well depend on crop diversity, the view that the two objectives of diversity and productivity are, in fact, a question of trade-offs, may be erroneous.

Debate and research on agricultural biodiversity have focused largely on centers of crop origin and domestication (the majority of which are located in developing countries) due to their importance for the conservation of crop genetic resources. However, the diversity of crop populations has economic importance in all production systems, regardless of whether they are predominantly characterized by modern varieties or by landraces. Concerns regarding the potentially negative effects of modern plant breeding technology on crop genetic diversity, despite successes in improving both food supply and food security, have been raised for both modern and traditional systems. These concerns have not always been supported by empirical evidence of increased genetic uniformity; research on diversity levels in recent time periods has demonstrated that levels of crop genetic diversity have been increasing in modern wheat systems (Smale et al. 2002; Reif et al. 2005; Warburton et al. 2006).

As a result of these ongoing debates, several questions reoccur: (1) How diverse are genetic resources?; (2) What is the value of genetic resources?; and (3) What can be done to ensure availability and access to crop genetic resources? (Day Rubenstein et al. 2005).

Economic Analysis of Crop Genetic Diversity

Economic analysis is well suited for such questions of value and behavior, and there is a growing body of literature addressing crop genetic diversity from an economic viewpoint. Economic analysis has already contributed to debates on the value of crop biodiversity, genetic resource conservation, both in situ and ex situ, exploration of short-term production trade-offs, and analysis of long-term stability and resilience (Smale 2006a; Smale 2005; Drucker and Smale 2005). The economic literature can be broadly categorized into research that has focused on: (a) the conservation of traditional crop varieties, including value to farmers and farmer incentives to continue their cultivation, and related institutional issues of seed systems and markets (Van Dusen and Taylor 2005; Smale et al. 2001; Meng et al. 1998; Brush et al. 1992); (b) the productivity of genetic resources and diversity, including ex situ conservation, returns to productivity research, and whether and how crop genetic diversity enhances the economic value of crop output (Koo et al. 2004; Smale and Koo 2003; Evenson and Gollin 2003; Evenson et al. 1998); and (c) broader issues of the value of biodiversity such as the costs and benefits of preventing the extinction of species that have aesthetic, intrinsic, or indirect use value to humans through supporting the ecosystem in which they live (Costanza et al. 1997; Swanson 1995; Pearce and Moran 1994; Orians et al. 1990; Brown 1990).

The modeling of farmer decisions and incentives has generally been associated with the conservation of traditional crop varieties, whereas productivity-related research has generally been associated with modern varieties. We suggest this division is artificial, given that many of the same issues and principles apply to both traditional and modern systems. Research efforts continue to clarify the relationship between crop diversity and the principal actors involved in the use of crop genetic resources, whether they be farmers or scientists. The interaction between crop diversity and the welfare of farmers, many of whom live in subsistence environments, as well as the incentives for farmers to continue to cultivate a diverse pool of genetic materials, remain topics of particular interest (Smale 2006b). The examination of the influence of economic change on crop diversity and its determinants has also been extended to a wider range of crops and production and institutional settings (Smale 2006a).

In all of the research areas, policy environment and the role of policy are crucial. The determination and funding of agricultural research priorities and the structure of marketing and seed distribution systems are examples of the avenues through which policy can impinge on crop genetic diversity. An improved understanding of factors that affect the behavior of farm households and other decision makers will better inform the policies and practices of governments and national and international organizations to ensure conservation and to enhance the availability and use of agricultural biodiversity. Drucker and Smale (2005) emphasize that the economic analysis of agricultural biodiversity is not merely an intellectual exercise and that resulting knowledge should be applied to tools that support the decision-making process and used to inform policy decisions.

The integration of biophysical and ecological concepts with economic models has not been without difficulties. Several factors contribute to the challenges faced in empirical research, including multiple, not always consistent, definitions of diversity. The term "crop diversity"[1] encompasses many different aspects of diversity, ranging from the diversity maintained in a gene bank or in a breeder's working collection to that found in a farmer's field or in a region of commercial activity. Each is a distinct type of crop diversity with its own set of influencing factors, and economic models need to provide linkages between the diversity outcomes and economic behavior. Moreover, each of these levels of analysis of diversity can be examined through the application of various broad concepts of diversity. The concepts that have been most utilized in economic applications are those that measure diversity over space and time (spatial diversity and temporal diversity); those that classify diversity based on whether or not it is visible (apparent diversity and latent diversity); and those that distinguish variation among crop varieties (inter-varietal diversity) versus diversity within a crop variety (infra-varietal diversity).

[1]*Agricultural biodiversity refers to all diversity within and among species found in domesticated crop systems while crop biodiversity (or diversity) refers to the biological diversity of crops, including varieties recognized as agro-morphologically distinct by farmers and genetically distinct by plant breeders (Smale 2005).*

Appropriate data for the representation of diversity and measurement issues also need to be identified. A crucial factor is a clear definition of the crop population being examined, since different users of crop diversity (e.g., farmers, plant breeders, gene bank curators) will often use their own taxonomies or systems of distinguishing among plants. The choice of the most appropriate metric to represent a concept of diversity, usually some kind of index, also needs to be considered in the context of the proposed analysis. As with the taxonomy, a diversity metric may also implicitly incorporate the priorities of the groups that utilize it. For example, breeders and geneticists are likely to prefer metrics that weight more heavily criteria identifying the most promising materials for crop improvement (Smale 2006b), such as traits of interest related to breeding priorities and the likelihood of heritability. The genetic resources identified by these metrics may or may not coincide with the materials valued by farmers.

Finally, specifying and testing economic relationships between diversity outcomes and the decisions of economic actors, such as farmers in their fields, can be more straightforward for certain concepts of diversity than others. Methodological progress has been considerable. It is clear that the choice of diversity concept and metric must have some logical and explainable association with the scope and unit of analysis. Relatively more complex representations of diversity, often unobservable to farmers and other decision makers, are certainly useful for some purposes, but may not be the most appropriate to include in a farm level analysis (Smale 2006b; Meng et al. 1998). Recent research also suggests the need for improved understanding of the linkages between diversity at different levels of analysis, for example, between household and village or provincial levels in order to better identify the most efficient point of entry of a policy or program intervention (Smale et al. 2006).

An Overall Conceptual Framework

A clearer elucidation of an overall framework in which relevant components of the economic analysis of crop diversity can be placed could contribute considerably to providing an idea of the current state of play. One of the principal motivations for the Australian Centre for International Agricultural Research (ACIAR) project, from which stemmed much of the research reported in this volume, was a recognition of the usefulness of such a framework. What was envisioned was not a rigid framework, but rather something that would allow researchers to identify patterns and points of linkages among the existing literature. Moreover, given the proliferation of diversity concepts and diversity indices, there is a growing need to take stock of what has been undertaken and to continue the discussion of the appropriate use of diversity concepts and indicators for different research objectives. We need to look at where we are now in the discussion of whether there is a "best" concept of diversity that should be associated with a certain type of economic analysis, or whether there is a "best" measure of diversity to represent a given diversity concept. Similarly, more clarification is needed on how much the available measures of diversity are substitutable, and if not, whether we need to be aware of any qualifications in using a "second best" concept or measure.

We envisioned a multidimensional framework including alternative concepts of diversity and encompassing various diversity concept and scope of analysis combinations, as well as continuing the examination of appropriate ways to measure and represent diversity levels. The general framework would address diversity at various units of observation, including field and household level as well as higher levels of aggregation. Ideally, it would also integrate research issues involving in situ and ex situ conservation methods, temporal and spatial differences in diversity, and impacts on production and consumption. Theoretical relationships between observed crop diversity and the human and agro-ecological factors specific to different levels of analysis would also be considered. More generally, such a framework could perhaps provide an organizational foundation for future research in the economics of agricultural biodiversity.

In this volume, we present an attempt at assembling this overall conceptual framework, as well as formulating and applying a subset of methodologies that enable economic researchers to address issues related to crop diversity. The range in the level of analysis, from household to national level, permits reflection on linkages between farm level studies and higher geographical levels of scale as well as on the implications of policy interventions at different levels. Our general approach has been to define and develop building blocks, or key inputs, that can be assembled for analytical use in several different ways. The first essential building block continues the work of Meng et al. (1998, 1999) and Smale et al. (2003) in selecting and adapting scientific concepts and indices of diversity for applied economic use. We place these concepts and indices in the appropriate part of our conceptual framework where their use corresponds with the type of economic analyses undertaken and discuss the data requirements for their use. The second building block continues research that identifies determinants of diversity—factors that influence the supply of diversity and diversity outcomes—and includes specification and testing of economic models of crop diversity outcomes at different levels of analysis. A third building block examines the economic impacts of crop diversity through possible effects on crop productivity.

We illustrate selected parts of the overall framework with wheat production and diversity data from Australia and China. Wheat has a long history of scientific improvement, and although landraces, or traditional varieties, continue to be cultivated, the large majority of wheat area is cultivated in improved varieties (Lantican et al. 2005; Heisey et al. 2002). Globally, wheat is one of the most significant crops, with total production in 2005 reaching 623 million tons. China is the largest producing country, with close to 16% of world production in 2005, and an intermittently significant wheat importer. Australia is also a significant producer with almost 4% of world production in 2005, as well as an important player in international wheat trade. In 2003, wheat supplied a world-wide average of 513 calories per head per day, or approximately 18% of total calorie intake, thus providing a significant food source on a global basis (FAOSTAT 2007).

The focus of the applied research presented in this volume addresses diversity in wheat production systems characterized completely or to a large extent by the cultivation of modern varieties. The detailed case studies contribute to the economic analysis of crop diversity for a range of modern production environments and socioeconomic conditions. Diversity among different varieties of wheat, or inter-varietal diversity, is more relevant in these environments than the infra-crop diversity more often observed in areas where traditional varieties or landraces are cultivated, where diversity is an outcome of both natural selection and farmer selection and management.[2] While infra-varietal diversity, an important element of landrace diversity, and diversity among crops are both included in the overall conceptual framework, specific analyses related to the in situ conservation of crop diversity in traditional varieties and to diversity among crops are not addressed here. Also not explicitly addressed in the research in this volume are issues related to the costs and benefits of the conservation of crop genetic resources, whether ex situ or in situ.

Why China and Australia?

With differing degrees of commercialization, breeding and research policies, and incentives for household variety choice, the wheat production systems of China and Australia provide interesting material for comparison. Previous work on wheat diversity in Australia provides a foundation from which to move forward (Brennan 1989; Brennan and Fox 1998). Although aggregate level data on diversity for China and Australia are not identical, they are similar enough to permit comparisons between wheat production systems in the two countries. The case studies from Australia and China provide the structure for examination of relationships among various concepts of diversity used for analysis at aggregate and household levels. Not all elements of the framework can be addressed, but our data allow us to examine temporal and spatial diversity, apparent and latent diversity, and inter-varietal diversity at varying levels of analysis and aggregation in China and Australia.

[2] *In the case of on-farm landrace cultivation, the role of landraces as global public goods and implications of diversity outcomes on the conservation of crop genetic resources present an additional factor to consider. Furthermore, because most centers of origin and domestication where landrace cultivation is observed are located in developing countries, perceived trade-offs between diversity and production are complicated by the added dimension of possible trade-offs between rural development and conservation (Smale et al. 2006).*

The Australian analyses focus on the impacts on diversity outcomes at aggregate levels, including the national, state, and shire levels, of industry changes and policy changes in prioritizing and funding research. Changes in the wheat marketing system and in the commercialization of research are also examined in the context of wheat diversity outcomes. Additionally, information from a breeders' survey conducted in Australia enables a better understanding of the link between some of these policy changes and variety supply, a significant determinant in observed diversity outcomes.

The Chinese applications examine changes in wheat diversity over time at the national and provincial levels, an analysis not previously possible due to lack of required data. Another useful component of the Chinese research includes an analysis of household varietal choice decisions. Through survey data collected at the household level in China, the role of both production- and consumption-related factors on varietal choice at the farm level are explored. A more complete understanding of farmer choice is useful in identifying important traits for future breeding research as well as exogenous market-related and policy factors that influence household behavior and diversity outcomes. The Chinese household study also provides an interesting contrast on farmer incentives in an area of modern variety cultivation relative to existing studies in areas of crop domestication and diversity.

Organization of this Book

This book has four major themes: first, a brief examination of the role played by economic and other policies in influencing levels of wheat genetic diversity; second, the continuous changes taking place over time in wheat genetic diversity in the context of changing policy backgrounds and priorities; third, the identification of factors that explain changes and differences in spatial diversity; and finally, an analysis of the productivity impacts of those changes.

Prior to exploring the themes in detail, however, the concepts and means of measurement of crop diversity are defined and placed into the appropriate context. In Chapter 2, we expand upon our discussion of a framework for economic analysis of crop diversity and review various concepts and measures of diversity. We discuss differences in the meaning of the measures and the likely areas in which each will be useful are explained, as well as present the ones that will be utilized throughout this book.

In Chapter 3, we begin to focus on the policy context for crop diversity and the ways in which policies influence the supply of crop diversity. The historical policy influences on genetic diversity changes in both China and Australia are then explored. The potential impact pathways of government policy change on the supply and demand for diversity in wheat production are diverse and often indirect. However, it is clear that policy changes, both at the macroeconomic level and at the industry level, can have a significant impact on the availability of genetic diversity and its usage on farms, both in

China and Australia. The differing policy frameworks and institutional settings for the two countries over recent decades provide a valuable contrast.

Breeders' demand for and usage of wheat genetic resources are explored from an Australian viewpoint to provide a better understanding of the linkages between policy changes and supply of varieties. Key issues are the breeders' need for genetic diversity in their crossing populations and the availability of appropriate material that can be readily incorporated into those programs. Another influential issue identified that affects variety supply is the funding environment and the extent to which such an environment can impinge on the breeders' enhancement of genetic diversity in their gene pools.

In Chapter 6 and Chapter 7, changes in wheat diversity levels are explored empirically for both China and Australia. The considerable changes that have occurred in China since 1978 are explored in detail using concepts and indices introduced in Chapter 2. Similarly, changes that have occurred in Australia since 1965 are documented. In each case, the differences in diversity using measures applied at varying levels of aggregation (e.g., national, province/state, regional) are examined.

The research described in Chapter 8 and Chapter 9 addresses the factors leading to changes in spatial diversity in both Australia and China. Two different analyses are presented. The first uses data from both countries to identify common elements that can explain changes in wheat diversity outcomes on farms. For Australia, the level of analysis takes place at the shire level in New South Wales, while in China, the analysis is carried out at the provincial level. The modeling of variety area shares permits the analysis of policies that promote more favorable spatial distribution of varieties. The role of specific policy and environmental factors, as well as the pivotal role of the supply of diversity through plant breeding programs, can be more clearly observed. A more disaggregated analysis of the determinants of wheat diversity in Chinese household farms is also carried out. This study explores influential factors, related to both production and consumption, for varietal choice and diversity outcomes at the household level. Since, ultimately, diversity outcomes are determined at the farm level, the identification of significant determinants of farm level decisions is a crucial part of diversity research.

In Chapter 10 and Chapter 11, the relationship between productivity and diversity in China is explored in two different ways. An analysis using total factor productivity is applied to China at both the national and provincial levels to test the significance of the relationship between wheat diversity and productivity. An alternative cost function approach is also applied to Chinese data at the provincial level. The cost function analysis explicitly including prices and other economic variables highlights economic efficiency and permits the examination of changes in the cost of wheat production associated with an increase in measured levels of crop diversity or a "marginal cost of diversity." Parallel productivity analyses for Australia were not possible due to the lack of required data.

Lastly, in Chapter 12, we present an overview of the key findings from the application of economic analysis of diversity in China and Australia. We then highlight implications for policy and examine some of the methodological issues requiring further analysis.

REFERENCES

Brennan, J.P. 1989. Wheat varietal turnover and diversification in New South Wales. *Agricultural Science* 2: 27-31.

Brennan, J.P. and P.N. Fox. 1998. Impact of CIMMYT varieties on the genetic diversity of wheat in Australia, 1973-1993. *Australian Journal of Agricultural Research* 49: 175-178.

Brown, G.M. 1990. Valuing Genetic Resources. In *Preservation and Valuation of Biological Resources.* G.H. Orians, M. Brown, W.E. Kunin and J.E. Swierzbinksi (eds). Seattle: University of Washington Press.

Brush, S.B., J.E. Taylor, and M.R. Bellon. 1992. Technology adoption and biological diversity in Andean potato agriculture. *Journal of Development Economics* 39: 365-387.

Cassman, K.G., S. Wood, P.S. Choo, C. Devendra, J. Dixon, J. Gaskell, S. Khan, R. Lal, J. Pretty, J. Primavera, N. Ramankutty, E. Viglizzo, D. Cooper, S. Kadungure, N. Kanbar, S. Porter, and R. Tharme. 2005. Cultivated Systems. In *Millennium Ecosystem Assessment, Ecosystem and Human Well-being, Vol 1: Current Conditions and Trends.* R. Scholes, R. Hassan and N. Ash (eds.). Washington D.C.: Island Press.

Costanza, R., R. d'Arge, R. de Groot, S. Farber, M. Grasso, B. Hannon, K. Limburg, S. Naeem, R. O'Neill, J. Paruelo, R. Raskin, P. Sutton, and M. van den Belt. 1997. The Value of the World's Ecosystem Services and Natural Capital. *Nature* 387: 253-259.

Cox, T.S., J.P. Murphy, and M.M. Goodman. 1988. The contribution of exotic germplasm to American agriculture. In *Seeds and Sovereignty: The Use and Control of Plant Genetic Resources.* Durham N.C.: Kloppenburg, J.R. Jr. (ed.). Duke University Press.

Day Rubenstein, K., P. Heisey, R. Shoemaker, J. Sullivan, and G. Frisvold. 2005. *Crop Genetic Resources: An Economic Appraisal.* Economic Information Bulletin, Number 2. Washington D.C.: USDA/Economic Research Service.

Drucker, A. and M. Smale. 2005. Chapter I: Overview. In *The State of Economics Research: Sustainable Management of Crop and Livestock Biodiversity.* Report prepared for the CGIAR System-wide Genetics Resources Program.

Evenson, R.E. and D. Gollin (eds.) 2003. *Crop Variety Improvement and its Effect on Productivity: The Impact of International Agricultural Research.* Wallingford, UK: CABI Publishing.

Evenson, R.E., D. Gollin, and V. Santaniello (eds). 1998. *Agricultural Values of Plant Genetic Resources.* Wallingford, UK: CABI Publishing.

FAOSTAT 2007. Rome: FAO.

Fowler, C., M. Smale, and S. Gaiji. 2001. Unequal exchange? Recent transfers of agricultural resources and their implications for developing countries. *Development Policy Review* 19: 181-205.

Frankel, O.H. 1970. The genetic dangers of the Green Revolution. *World Agriculture* 19: 9-13.

Harlan, J.R. 1972. Genetics of disaster. *Journal of Environmental Quality* 1: 212-215.

Heisey, P.W., M.A. Lantican, and H.J. Dubin. 2002. *Impacts of International Wheat Breeding Research in Developing Countries, 1966-97*. Mexico, D.F.: CIMMYT.

Hoisington, D., M. Khairallah, T. Reeves, J.-M. Ribaut, B. Skovmand, S. Taba, and M. Warburton. 1999. Plant genetic resources: What can they contribute toward increased crop productivity? *Proceedings of the National Academy of Science* 96: 5937-5943.

Jana, S. 1999. Some recent issues on the conservation of crop genetic resources in developing countries. *Genome* 42: 562-569.

Koo, B., P.G. Pardey, and B.D. Wright. 2004. *Saving Seeds: The Economics of Conserving Crop Genetic Resources ex situ in the Future Harvest Centres of the CGIAR*. Wallingford, UK: CABI Publishing.

Lantican, M.A., H.J. Dubin, and M.L. Morris. 2005. *Impacts of International Wheat Breeding Research in the Developing World, 1982-2002*. Mexico, D.F.: CIMMYT.

Meng, E., M. Smale, H., Ruifa, J.P. Brennan, and D. Godden. 1999. Measuring crop genetic diversity for economic analysis: Concepts, indices and applications. Contributed paper, 43nd Annual Conference of the Australian Agricultural and Resource Economic Society, Christchurch, New Zealand.

Meng, E.C.H., M. Smale, M.R. Bellon, and D. Grimanelli. 1998. Definition and measurements of crop diversity for economic analysis. In *Farmers, Gene Banks and Crop Breeding: Economic Analyses of Diversity in Wheat, Maize, and Rice*. Smale, M. (ed.). Boston: Kluwer Academic Publishers.

Orians, G.H., G.M. Brown Jr., W.E. Kunin, and J.E. Swierbinski (eds.). 1990. *The Preservation and Valuation of Biological Resources*. Seattle: University of Washington Press.

Pearce, D. and D. Moran. 1994. *The Economic Value of Biodiversity*. London: Earthscan Press.

Reif, J.C., P. Zhang, S. Dreisigacker, M.L. Warburton, M. van Ginkel, D. Hoisington, M. Bohn, and A.E. Melchinger. 2005. Wheat genetic diversity trends during domestication and breeding. *Theoretical and Applied Genetics* 110: 859-864.

Singh, R.P., D.P. Hodson, Y. Jin, J. Huerta-Espino, M.G. Kinyua, R. Wanyera, Smale, M. 2005. Chapter II: Economics literature about crop biodiversity: Findings, methods and limitations. In *The State of Economics Research: Sustainable Management of Crop and Livestock Biodiversity*. Report prepared for the CGIAR System-wide Genetics Resources Program.

Smale, M. (ed.). 2006a. *Valuing Crop Diversity: On-farm Genetic Resources and Economic Change*. Wallingford, UK: CABI Publishing.

Smale, M. 2006b. Concepts, metrics and plan of the book. In *Valuing Crop Diversity: On-farm Genetic Resources and Economic Change*. Smale, M. (ed.). Wallingford, UK: CABI Publishing.

Smale, M. and B. Koo (eds.) 2003. Genetic Resources Policies: What is a Gene Bank Worth? Research at a Glance. Briefs 7-12. International Food Policy Research Institute (IFPRI), Washington, D.C., International Plant Genetic Resources Institute (IPGRI) and Systemwide Genetic Resources Program (SGRP), Rome.

Smale, M., M.R. Bellon, and J.A. Aguirre Gomez. 2001. Maize diversity, variety attributes, and farmers' choices in Southeastern Guanajuato. *Economic Development and Cultural Change* 50: 201-225.

Smale, M., M.P. Reynolds, M. Warburton, B. Skovmand, R. Trethowan, R.P. Singh, I. Ortiz-Monasterio, and J. Crossa. 2002. Dimensions of Diversity in Modern Spring Bread Wheat in Developing Countries from 1965. *Crop Science* 42: 1766-1779.

Smale, M., E. Meng, J.P. Brennan, and R. Hu. 2003. Determinants of spatial diversity in modern wheat: Examples from Australia and China. *Agricultural Economics* 28: 13-26.

Smale, M., L. Lipper, and P. Koundouri. 2006. Scope, limitations and future directions. In *Valuing Crop Diversity: On-farm Genetic Resources and Economic Change.* Smale, M. (ed.). Wallingford, UK: CABI Publishing.

Swanson, T. 1995. *The Economics and Ecology of Biodiversity Decline*. Cambridge: Cambridge University Press.

Van Dusen, M.E. and J.E. Taylor. 2005. Missing markets and crop diversity: Evidence from Mexico. *Environment and Development Economics* 10: 513-531.

Warburton, M.L., J. Crossa, J. Franco, M. Kazi, R. Trethowan, S. Rajaram, W. Pfeiffer, P. Zhang, S. Dreisigacker, and M. van Ginkel. 2006. Bringing wild relatives back into the family: recovering genetic diversity in CIMMYT improved wheat germplasm. *Euphytica* 3: 289-301.

Chapter 2

Conceptual Framework for Crop Diversity Concepts and Measurement

E. Meng, M. Smale, and J.P. Brennan

ABSTRACT

A number of concepts and tools have been developed to measure and classify crop diversity. Here we synthesize these concepts and measures to enable the development of a framework within which each can be considered and analyzed at different levels (farm, district, region, state/province, national). The concepts of spatial and temporal diversity and apparent and latent diversity are incorporated into the framework. We then identify different indices to represent each concept, explain the application of the diversity concepts and indices used for the analyses in this volume, and draw implications for economic analysis and the choice of diversity measure.

The applications of economic analysis to issues of crop diversity are numerous. They include questions of priority setting and resource allocation, conservation, property rights, contributions to productivity and stability, and valuation. Economic models have addressed the cropping decisions made by household farms, the choice of germplasm materials by breeders and gene bank curators, as well as the implications of research, development, and trade policies. In the context of this economic research, agricultural biodiversity has been examined in focused studies ranging from the variation within single crop populations in a farmer's field to the broader setting of a cropping system. All are examples of valid, albeit different, approaches to defining crop diversity. Differences in the historical evolution of a crop, its interactions with agro-ecological conditions, and how it is used by producers and consumers are also reflected in different ways in crop populations. Moreover, the analytical approach and implications for the context of a traditional crop population or set of populations can be quite different from those that might apply in the case of a modern crop population. The existence of this range of contexts underlines the importance of the potentially quite different implications of diversity in selecting appropriate approaches for analysis and in interpretation.

This chapter addresses issues to consider when incorporating crop diversity into economic analyses, presenting a synthesis of several relevant concepts and tools, and outlining a framework within which these concepts and measures can be organized and compared. We conclude the chapter with a discussion of implications for economic analysis.

Ecologists, crop scientists, and geneticists have utilized a range of diversity estimators and methodologies to classify diversity (e.g., Hawksworth 1995; Magurran 1988), many of which may be adapted for economic analysis of crop genetic diversity. Biological diversity refers in general to a broad area of scientific inquiry encompassing all living organisms and their relationship to each other. The abundant literature on biodiversity, and more specifically agricultural biodiversity, can generate some confusion over the definition, measurement, and interpretation of crop diversity in the context of economic analysis. For applied economists, one major dilemma is how best to relate diversity concepts defined for biological and genetic phenomena to the economic decisions of farmers and other decision makers in a way that is meaningful for economic analysis. Diversity indices that are complex in terms of data and tools can often be difficult to link to decisions made at the farm level (Smale 2006). However, when measuring diversity levels in a collection of breeding materials that reflect the decisions of plant breeders, such diversity indices are likely to make more sense.

Three primary factors play a key role in considering an analytical framework: (1) the level or unit of analysis; i.e., who is making the decisions and what is the relationship between the decision maker and crop diversity; (2) the concept of diversity, which will be determined largely by the level and unit of analysis; and (3) the representation of diversity; i.e., what kind of data will be used to construct the diversity measure and what kind of metric, all of which are also closely linked to the level and unit of analysis.

Levels of Economic Analysis

Farm or household level studies have largely focused on traditional crop populations in environments characterized by subsistence or semi-subsistence households for whom production and consumption characteristics are important (e.g., see Van Dusen and Taylor 2005; Smale et al. 2001; Meng et al. 1998; Brush et al. 1992). The decisions of these farmers are private decisions, whether modeled at the plot or household level, but carry particular implications for conservation of crop genetic resources of global interest. Significant diversity can potentially arise from variation both within and among varieties of a specific crop as well as among different crops. The examination of household decisions relevant to diversity could thus include the decision to cultivate a specific crop variety, a set of crop varieties, or diversification across crops. In considering and specifying possible relationships between household and diversity outcomes, the nature of the particular crop varieties or crops of interest need to be considered, since management and diversity relationships differ for open-pollinated and cross-pollinated crops, as well as for vegetatively propagated crops. Increasingly, it is not only the cultivation decisions of the household that are of interest (e.g., decisions on whether or not to cultivate a traditional variety within one or multiple crops and the resulting diversity outcome), but also the household's management of the varieties and seed as well as their interactions with seed systems and markets. Welfare implications for households maintaining high on-farm levels of diversity are also being explored, as well as the role of interventions to improve welfare for these households while maintaining diversity.

Farm level applications are not limited to variety or crop decisions concerning traditional varieties. Crop diversity also exists outside centers of crop diversity and domestication. In many situations, modern varieties fulfill a household's multiple production and consumption objectives in a way commonly associated with traditional varieties. In more commercialized production environments, crop diversity outcomes are also of interest for households, particularly as they relate to perceived trade-offs and decisions among household diversity outcomes, productivity, and household welfare. Regardless of the crop or degree of commercialization, taxonomies used to identify crop populations at the household level have most commonly been based on names or characteristics of the crop and crop varieties that are observable or otherwise distinguishable (i.e., through touch or taste) to household members—rather than diversity, say, at the molecular level.

Crop scientists perceive and utilize crop diversity in ways different from those of farmers. Crop diversity from a breeder's perspective is a breeding tool, as well as potentially one of many breeding objectives necessary to balance in the development of new varieties. Incorporation and expansion of novel sources of diversity in a breeding program, whether it be an immediate product-oriented goal or a more general, long-term one, is a conscious objective for the breeder in a way that is different from the relationship between farmers and the farm level diversity they maintain. Similarly, from the perspective of a gene bank manager, the maintenance of certain levels of diversity within a collection is a means to ensure the availability of an adequate representation of existing diversity in gene bank collections and the prevention of genetic drift. Distinguishing within and classifying phenotypic variation, including the use of agro-morphological characteristics important to farmers, is one approach to ensure and distinguish differences in crop populations. Information on agronomic and morphological traits is also used by gene bank managers to classify materials and by crop breeders to enable the introduction of novel variation for selected traits into the breeding population (Lage et al. 2003). Underlying the use of observable characteristics is the assumption that they are based on a limited number of genes associated with the expression of certain traits that are selected for by breeders or farmers or through a process of natural selection.

Molecular-based methods to discern and manage genotypic variation have also become increasingly important for representing diversity levels, particularly as the costs of such methods decrease. For scientists concerned with incorporating or confirming diversity in germplasm collections, molecular-based diversity measures are meaningful and useful (Smale et al. 2002; Dreisigacker et al. 2004). Moreover, as molecular markers are increasingly utilized as a crop breeding tool, it makes sense to use data from molecular-based diversity studies.

Analyses of crop diversity at levels of aggregation above the household have implicitly acknowledged the primary role of farmers in determining the micro-level diversity outcomes that shape aggregate diversity outcomes. These analyses focus primarily on understanding the parallel but larger-scale relationships between diversity outcomes and agro-ecological conditions, infrastructural and institutional factors, and policy regimes. Aggregate analyses have included diversity based on named varieties, morphological characteristics, variety pedigrees, and molecular data. The use of this range of taxonomies is both more ad hoc—there are no preconceived relationships between aggregate level decisions and diversity outcomes per se—and justified, since all diversity outcomes are valid and apply at aggregate levels. A relevant question is the extent to which the diversity represented by different taxonomies is similar at aggregate levels. Also, it is not clear in what situations there may be conflicting pressures on diversity levels as represented by different concepts of diversity from significant factors. Smale (2006) notes that there were no apparent tradeoffs conservation objectives among different subcategories of spatial diversity in recent farm level studies on individual crops, although methodological limitations may have impaired the ability to observe significant differences.

The linkages between household level analysis and analysis at higher levels of aggregation have not been extensively examined in the economic literature, largely due to data limitations. To a certain extent the nature of the crop population in question will drive the relationship and its implications. The link between household level choices and aggregate level diversity outcomes in the context of traditional crop varieties carries consequences for on-farm conservation of unique crop genetic resources of global significance in a way not relevant for modern varieties. Linkages in a setting of modern crop varieties have potentially significant, but different implications for issues of crop uniformity and productivity.

Diversity Concepts in Economic Analysis

The appropriateness of the diversity concept is important to consider vis-á-vis the objectives of the study and of the level at which the analysis takes place

(Meng et al. 1998). A given diversity concept can be expressed over a range of different taxonomies; that is, crop populations can be classified by the names or criteria that farmers use to describe them, by their genealogies as recorded by plant breeders, or by the genetic identity that molecular analysis reveals. The distinction between a concept of diversity and its representation with an index or some other measurement tool and appropriate data is worth reemphasizing. As a mathematical construct, the index incorporates information to express the concept of diversity as a number or scalar. The index is indeed often linked to the concept, but the analytical framework and diversity concept drive the decision of what index to use, not vice versa. Various concepts of diversity used in economic analysis have been described previously in several studies (Meng et al. 1998, 1999; Smale 2006; Day Rubenstein et al. 2005). Here we briefly review some of the most relevant concepts and discuss their use at different levels of analysis.

Spatial diversity—the amount of diversity in a given geographical area—is the most commonly used concept of diversity in recent economic analyses of crop diversity. Magurran (1991) classifies ecological indices of species diversity by three criteria: (1) species richness, or the number of species encountered in a given sampling effort; (2) relative abundance, or the number of individuals associated with each of the species; and (3) evenness of species or proportional abundance. A count of species reported or collected in a defined area, although usually simplest to implement, assumes that all species at a site contribute equally to its biodiversity (Harper and Hawksworth 1995). Since this may not be the case, frequency counts of individuals within a species (relative abundance or inverse dominance) provide more information on whether or not certain varieties or groups of varieties dominate others. The third category, evenness, combines a measure of proportional representation with the number of species. Also called "equitability," it refers to the degree of equality in the abundance of the individuals, or the relative uniformity of their distribution across species. When all species in a sample are equally abundant, evenness reaches a maximum (Ludwig and Reynolds 1988).

Spatial diversity has been represented in economic research at both household and aggregate levels of analysis for diversity outcomes within an individual variety or crop population, among varieties of the same crop, and across multiple crop species. Applications of spatial diversity concepts have focused on diversity outcomes resulting from farmer decisions in centers of domestication and diversity that are of interest for the conservation of crop genetic resources and farmer valuation of crop diversity. More recently, they have been utilized to examine diversity outcomes in more commercially-oriented production systems where issues related to genetic uniformity have presented concerns to productivity. Studies in Australia and the United States as well as the aggregate studies on Australia and China in this volume are examples.

For a traditional variety, whether it be an open-pollinating or self-pollinating crop, within a field both morphological and molecular diversity are likely to be found, since landraces are not uniform and, indeed, there may be little incentive for uniformity. In some cases, the level of variability present in one landrace might also satisfy a household's demand for production stability, a guarantee of at least something in an uncertain environment of biotic and abiotic stresses. However, given the multiple household demands for production and end-use characteristics that need to be satisfied in a traditional system, farmers often choose to grow more than one variety. In traditional systems, crop diversity is likely to be significant as a result not only of the variation existing among crop varieties, but also of variation within an individual traditional variety or landrace. In general, diversity for self-pollinated crops such as wheat is expected to be distributed among rather than within varieties, with the opposite holding for open-pollinated crops such as maize. However, it is the relative amounts of diversity among varieties and within a variety, rather than absolute levels, that are of more importance, particularly for policy considerations.

In a commercial field, a modern variety cultivated is likely to exhibit a higher level of genetic and morphological uniformity. In that situation, farmers in effect choose diversity within a crop by planting a number of different varieties in different fields, although as named varieties can often be closely related genetically, the cultivation of multiple varieties does not guarantee genetic diversity.

However, as geographical aggregation increases (from a field to a farm to a district, for example), the number of decision makers increases as well as the number of different production environments. Thus, in a commercial system, while each individual farmer may plant only one or two varieties, across the entire district the choices of variety will differ with farmer preferences and soil changes, even within the same climatic region. The greater the extent of the aggregation, the more the total range of cultivated varieties is likely to increase. As aggregation occurs across production environments (including soils and climate), it is likely that measured diversity will increase, as different varieties are likely to be grown where there is a diverse set of production environments with differing production constraints. The consideration of relevant taxonomy may need to be revisited, as a taxonomy appropriate at one level of analysis may change as the level of aggregation increases.

Changes over time in spatial diversity can be viewed as a kind of temporal diversity, although temporal diversity in previous research has been commonly associated with the rate of turnover of cultivated varieties over time (Brennan and Byerlee 1991). Jin et al. (2002) also define a measure of varietal turnover used in Chapter 10 that is based on the extent to which newly introduced varieties replace existing varieties. In addition to the idea of temporal diversity as a necessary substitute for spatial diversity, particularly in modern agricultural systems (Duvick 1984), temporal diversity is perceived to be more a measure of the output and success of crop breeding programs and agricultural research systems, due to their role in the development and dissemination of varieties and seed (Day Rubenstein et al. 2005). Temporal diversity has been studied commonly in the context of commercial production systems, but also applied to analysis in developing countries. The replacement of varieties in regular varietal turnover capitalizes on advances in knowledge and technology and reduces potential risks of disease that result from pathogen mutations that have overcome the genetic resistance in older varieties.

Crop diversity is not necessarily readily observable, and the concepts of apparent and latent diversity reflect that reality. Apparent variation refers to the physical variation in traits that can be observed by farmers or scientists in the field (Meng et al. 1998). These characteristics may include concrete distinctions such as height or color, but can also refer to more complex characteristics such as observable yield potential, yield stability, and heat and drought tolerance (Smale et al. 2002). Latent diversity has been defined by Souza et al. (1994) as that diversity which is not observable until "stimulated" to expression by a new pathogen or other changes in the growing environment. It has generally been represented by diversity in breeding materials used as parents and through variation at the molecular level. Although they are separate diversity concepts per se, these concepts and those of spatial diversity or temporal diversity are not mutually exclusive. Both spatial and temporal diversity can be either apparent or latent, although the concepts of temporal diversity utilized thus far are largely latent, due to the strong association between the concept of temporal diversity and varietal turnover. Nevertheless, it may be more logical to consider apparent and latent diversity, in a sense, as subcategories of both spatial and temporal diversity.

Diversity Indices in Economic Analysis

The same index or mathematical construct can be applied using different systems of classification of a crop population, or taxonomies, to express a selected diversity concept. The use and interpretation of diversity indices requires caution, since diversity outcomes based on different taxonomies may require different interpretations or carry different implications. Data availability, as always, will also be a consideration.

For spatial diversity, named varieties have constituted a commonly used taxonomy at all levels of analysis (household, aggregate) and for both traditional and modern crop populations. The precautions that should be observed when using named varieties for classifying traditional crop populations have been previously discussed (Meng et al. 1998). Relying on named crop populations may overestimate diversity, where genetically similar populations are identified by different names, or underestimate diversity if those identified by the same name possess important underlying genetic differences. This concern is likely to be more significant in the case of traditional varieties, or landraces, than for improved varieties, since the same landrace can be known by several names across villages or regions and, conversely, landraces with the same name across different regions may not be identical. Even with modern varieties, farmers often develop their own nomenclature, and depending on the reproductive nature of the crop, the variety cultivated in a farmer's field may have evolved significantly from a variety of the same name grown from newly purchased seed. In general, genotypic variation is known with greater precision in modern crop populations than in traditional varieties, due to the methodologies by which the former are developed and the stipulated requirements of distinctness and uniformity for varietal release. But similarities in genetic structure for named varieties may still be masked, as in the case of different varieties bred from the same parents or from sister lines, where pedigree data are not immediately or easily accessible.

Descriptions and formulas for diversity indices commonly used in economic analysis have been presented in previous literature (Smale 2006; Meng et al. 1999; Smale et al. 2003). Using named varieties as the chosen taxonomy, a count of named varieties is the most straightforward diversity index. Two other indices measuring spatial diversity, the Shannon index, representing evenness or proportional abundance, and the Berger-Parker index, representing relative abundance or inverse dominance, have also been frequently applied in recent economic studies using named varieties.

Variation in plant characteristics and other types of physical descriptors can also serve as the basis of the taxonomy. Analysis based on the specific characteristics and performance of plant populations addresses many of the concerns expressed previously about over- or underestimating diversity when relying solely on variety names. The traits considered may include both quantitatively and qualitatively measured descriptors of the crop's morphology and agronomic performance. Observable variation in plant characteristics can result either from genetic differences or differences in the environment. Genotype-by-environment interactions need to be assessed under controlled experiments with a rigorous trial design and sampling methodology. By using data from experimental trials designed to minimize the effects of such interactions, the certainty that the observed variation in traits will reflect genetic differences is increased.

Spatial indices such as the Shannon index are used to represent diversity in morphological data, where the frequency of appearance of selected traits replaces area shares of named varieties (Meng 1997; Zanatta et al. 1996). Spatial diversity can also be represented by the amount of difference or similarity based on distance measured between some defined characteristics or parameters (see Weitzman 1992; Solow and Polasky 1994). Franco et al. (1998) combine the Gower distance measurement with the Ward clustering methodology to form groups based on similarities in morphological characteristics and to enable analysis of significant differences among clustered groups. The technique defines groups by minimizing the within-group variance and maximizing between-group variance. Both quantitative and qualitative morphological data can be incorporated into this type of analysis.

Unobservable, or latent, information can also be used for measures of spatial diversity. Data on variety ancestry or pedigrees have been used to estimate genetic variation across a set of crop varieties based on principles of genetics and related assumptions that specific genes or gene combinations will be inherited from parents by resulting crosses. Pedigree-based measures, including measures of pedigree complexity and area-weighted coefficients of parentage and diversity have been described in detail in Meng et al. (1998) and Day Rubenstein et al. (2005) and utilized as a measure of spatial diversity in Smale et al. (2002). The coefficient of parentage (COP) estimates the probability that a random allele taken from a random locus in a variety X is identical, by descent, to a random allele taken from the same locus in variety Y (Malecot 1948). Values range from 0 to 1, with higher values indicating greater relatedness (for historical development and other examples of this concept, see Malecot 1948; Kempthorne 1969; Souza et al. 1994; Cox et al. 1986, 1985). The related measure, the coefficient of diversity (COD), is equal to one minus the COP between any pair of varieties.

Particularly when weighted by the percent of area planted to a given variety, average CODs represent spatial diversity as well as latent diversity (Souza et al. 1994). Because pedigree-based analyses of diversity require information as detailed and complete as possible on the ancestry of each variety, their applicability is limited to modern varieties with recorded genealogies. A point of reference for interpreting the magnitude of the COD is the value of 0.4375 assigned to varieties bred from the same parents, or from sister lines. Moreover, the area-weighted COD must always be lower than or, at best, equal to the average COD (the latter occurs only where the percentage distribution of wheat varieties by area shares is perfectly uniform).

Molecular fingerprinting tools, including simple sequence repeats (SSRs) and amplified fragment length polymorphisms (AFLPs), are now routinely used in diversity studies to estimate relationships between crop lines and populations. Spatial indices, such as the Shannon index, are likewise used to represent diversity in the form of abundance and frequency of observed alleles. Distance metrics, including Rogers' genetic distance and modified Rogers' difference among pairs of genotypes or groups of germplasm, are also utilized to assess differences in diversity levels (Warburton et al. 2006; Reif et al. 2005; Lage et al. 2003).

Weighted average variety age (Brennan and Byerlee 1991) has been one of the primary indices of temporal diversity utilized. The key data required to calculate it are the names of the varieties available within a given geographical area and their release dates, as well as information over time on the extent of their cultivated area. For landraces sown in the geographical area, the absence of any official release dates must be considered in the calculation of an index reflecting temporal diversity. The reliance on variety names for this index raises again the potential problems described earlier. Instead of named varieties, information on the development and incorporation of certain characteristics, such as genes to provide resistance to a specific disease or pest, can be used. Compiling this type of data requires considerably more information than for named varieties.

Linkages Across Taxonomies and Indices

Integrating the information gathered from the use of different combinations of indices and taxonomies has presented many challenges. There remains scope for clarifying differences in diversity outcomes found across different indices representing the same diversity concept and using the same taxonomy. It is not immediately clear in what conditions an analysis using, for example, the Shannon index to represent evenness in spatial diversity would result in different conclusions and implications from an analysis using another index also representing evenness, or from the Berger Parker index representing dominance in spatial diversity. Furthermore, by applying diversity indices from the ecological literature we are able to examine differences in spatial distribution of diversity; however, the added value of quantifying these differences and what significant economic outcomes, if any, they might have are not entirely clear.

One additional complication is that of conflicting information across taxonomies. Many studies have attempted to incorporate information from both farmers and scientists in classification efforts (Edmeades et al. 2006; Gauchan et al. 2006; Meng 1997). Reasons for potential discrepancies between diversity outcomes based on named varieties and others have already been discussed. Differences in describing diversity outcomes using agro-morphological and molecular-based measures are also a possibility; in certain crop populations, the presence of morphological differences may mask the closeness of the actual genetic relationship (Dudley 1994). Studies in the crop science literature have demonstrated the absence of a significant correlation between diversity measurements based on agro-morphological data and those based on molecular data (Lage et al. 2003). This lack of significance in correlation is explained by the lack of breeder selection on non-visible molecular data in contrast to the visible agronomic traits that are subject to selection (Koebner et al. 2002; Donini et al. 2000). Diversity measures using molecular data are based on the use of large segments of the relevant chromosome(s) and include both genes and gene combinations that may not necessarily express identifiable associated traits. Diversity measures based on molecular diversity are likely to be less biased, in the sense that they are associated with general chromosome segments rather than a set of genes specifically associated with observable traits (Lage et al. 2003). However, precisely because morphological diversity is visible and associated with useful traits, it may make more sense to use observable traits for modeling certain farm-level decisions, such as farm and plot level variety choice.

Molecular-based measures may be more appropriate for an unbiased view of diversity, but to assess diversity for selected visible traits, a morphology/performance-based measure may be preferable. Nevertheless, given that they reflect different information and since it remains unclear how best to combine agro-morphological and molecular information into one single measurement, diversity measurements from both types of data are seen as complementary, and the use of both is recommended for diversity analyses used by crop scientists (Lage et al. 2003).

What conclusions can we draw regarding the application of diversity indices in economic analyses of crop diversity? It is clear that at the household level, simple indices based on farmer taxonomies make the most sense. These can be linked more directly to farmer behavior and decisions and also be more straightforward to interpret. However, Smale (2006) points out that the careful choice of a taxonomy understood by both farmers and scientists is a possible means of integration. Or, to the extent that diversity has been analysed using both a farmer taxonomy and a genetically-based taxonomy more familiar to scientists, linkages established between the two can enable the use of more sophisticated diversity measures.

Several economic studies have examined the impacts of policies on diversity outcomes, comparing diversity indices that represent evenness as opposed to dominance in spatial diversity (Benin et al. 2006; Gebremedhin et al. 2006; Edmeades et al. 2006; Gauchan et al. 2006). As summarized by Smale (2006), no evidence was found to support hypotheses that policies would result in trade-offs between evenness in diversity and relative abundance in diversity. These joint findings would suggest that either the policies examined thus far are too blunt to be able to have a discernable effect on the nuances of spatial diversity, that there are measurement issues in the diversity indices used, or perhaps that we are reasonably safe in assuming that the indices of spatial diversity tested reflect diversity outcomes accurately enough for the purposes of our economic analysis.

Using the discussion above as a foundation, we summarize levels of analysis, diversity concepts, and associated use of taxonomies and indices within a general framework presented in Table 2.1. The diversity measures are categorized as spatial (apparent and latent), and temporal (latent). The measures that can be applied at the plot or field, household, community, and region/country levels are illustrated, and the data needed and some commonly used indices for each type of diversity at different levels are shown.

Applications of Diversity Concepts and Indices in this Book

The indices used to express the concepts and illustrations in this volume are presented in Table 2.2. It should be kept in mind that we are comparing

Table 2.1 Classification of diversity measures and common applications.

Type of diversity	*Unit of analysis*	*Data required*	*Commonly used measures*	*Comments*
Spatial-apparent	Plot/field	Morphological characteristics	Shannon index Clustering or distance metrics	Intra-crop variation; particularly relevant for landraces. Also potentially relevant for modern varieties of cross-pollinated crops and areas where seed replacement rates are low.
	Household	Named varieties Morphological characteristics	Count index Shannon index Berger-Parker index Clustering or distance metrics	Intra-crop diversity; detectable only with morphological characteristics. Both intra-crop and inter-crop diversity possible at household level. Shannon index used with both morphological traits and names. Diversity among crops has also been analyzed at household level.
	Community	Named varieties Named varieties/area share Morphological characteristics	Count index Shannon index Berger-Parker index Clustering or distance metrics	
	Region/ Country	Named varieties Named varieties/area share Morphological characteristics	Count index Shannon index Berger-Parker index Clustering or distance metrics	

Table 2.1 Contd.

Type of diversity	*Unit of analysis*	*Data required*	*Commonly used measures*	*Comments*
Spatial-latent	Plot/field	Molecular data	Shannon index Distance metrics	Intra-crop variation; particularly relevant for landraces.
	Household	Molecular data	Shannon index Distance metrics	Pedigree data; not very meaningful at household level.
	Community	Molecular data Pedigree data	Shannon index Distance metrics Area-weighted coefficient of diversity Pedigree complexity metric	Pedigree analysis; only applicable to modern varieties.
	Region/ Country	Molecular data Pedigree data	Shannon index Distance metrics Area-weighted coefficient of diversity Pedigree complexity metric	Pedigree analysis; only applicable to modern varieties.
Temporal-latent	Plot/field			Variety turnover rate at plot/field level has little significance for diversity analysis; higher at household level.
	Household	Household named varieties, area share and turnover rate		Very few panel data sets of household variety choice exist.

Table 2.1 Contd.

Type of diversity	*Unit of analysis*	*Data required*	*Commonly used measures*	*Comments*
	Community	Named varieties/area share Specific genes/gene combinations within named variety and area share	Rate of varietal turnover	
	Region/ Country	Named varieties/area share Specific genes/gene combinations within named variety and associated area	Rate of varietal turnover	
Temporal-apparent		None yet defined		

several things; namely, measurements for one diversity concept using the same indices but different data sets, measurements for one diversity concept using different indices but the same data sets, and measurements for differing concepts. Table 2.2 lists each index used by its name, category among spatial indices, and mathematical construction, with an accompanying explanation.

We have adapted and applied several indices commonly used to represent spatial diversity to data on wheat populations in China and Australia. To calculate diversity indices for China, we use three different taxonomies (named wheat varieties, morphological groupings, and pedigrees) as the basis for calculating various indices representing apparent spatial diversity and latent spatial diversity at both the national and provincial levels. Indices for apparent spatial diversity at the household level are also constructed. For Australia, we base apparent spatial diversity measures on named varieties and latent diversity measures on pedigrees at the national, state, and shire levels. Table 2.3 summarizes the levels of analysis, taxonomies, and diversity indices used for both China and Australia.

Implications for Economic Analysis

Having laid out different concepts of diversity and methods to represent them, it is evident that a judicious choice of measurement tools is essential to applied economists conducting research on crop genetic diversity. Since by its definition, the diversity based on pedigrees or differences at the molecular level is not observable to farmers, it may not be appropriate to model latent diversity as an explicit choice variable in models of decision-making at the farm level or even in aggregate analysis of regional crop productivity and stability. Farmers select crop populations to cultivate based on characteristics or qualities that are observable rather than on genetic structure that is not. There is seldom a direct relationship between the presence of an individual gene and a specific, physical characteristic. Most economically important traits are determined by multiple genes, and the relationship is usually quite complicated and often not yet completely understood. Consequently, the linkage between the economic decisions of farmers and genetic diversity measured at the molecular level is not straightforward conceptually or empirically. The relationship between a specific ancestor and an observable physical characteristic is also more often than not unclear, so that the inclusion of pedigree-based measures of genetic diversity in production analyses is difficult. With the exception of traits such as semi-dwarf height or certain disease resistances that can be traced to specific parents, most characteristics are the result of the combination of genetic material contributed by multiple parents. However, these difficulties do not diminish the usefulness of indices of latent diversity for describing and comparing crop diversity levels in the field or in a breeding program.

Table 2.2 Definition of spatial diversity indices used in this volume.

Index	*Category*	*Mathematical construction*	*Explanation*	*Adaptation in this volume*
Margalef	Richness	$D = (S - 1)/\ln N$ $D \geq 0$	Number of species (S) recorded, corrected for the total number of individuals (N) summed over species	S = number of wheat varieties grown in a season N = total hectares of wheat in that season
Berger-Parker	Inverse dominance (Relative abundance)	$D = 1/(N_{max}/N)$ $D \geq 1$	Measures dominance of the abundant species; the more dominant the most abundant species, the lower the index value	Inverse of maximum area share occupied by any single wheat variety
Shannon	Evenness (Proportional abundance)	$D = -\Sigma p_i \ln p_i$ $D \geq 0$	Measures evenness of the distribution of the species where p_i is the proportion, or relative abundance, of a species	p_i = area share occupied by i^{th} variety
COP	na	$0\ D \geq 1$	Genetic relationship between two cultivars based on their genealogies	Pair-wise genealogical correlation between two named (improved) varieties
WCOP	na	$0\ D \geq 1$	COP weighted by a selected measure of cultivar significance	COP weighted by variety area shares
COD	na	1-COP $D \geq 1$	Genetic differences between two cultivars based on their genealogies	Pair-wise genealogical differences between two named (improved) varieties
WCOD	na	$0\ D \geq 1$	COD weighted by a selected measure of cultivar significance	COD weighted by variety area shares

Source: Adapted from Smale et al. (2001), Meng et al. (1999), Souza et al. (1994), Cox et al. (1986, 1985). Mathematical construction as defined by Magurran (1991).

Table 2.3 Spatial diversity indices for Australia and China.

Country	*Level of analysis*	*Taxonomy/data*	*Diversity Index*
Australia	Shire	Named varieties	Shannon
			Margalef
	State	Named varieties	Shannon
			Margalef
	Country	Named varieties	Shannon
			Margalef
			Berger-Parker
		Pedigree	COD
			WCOD
China	Household	Named varieties	Shannon
	Province	Named varieties	Shannon
			Margalef
			Berger-Parker
			Number of varieties
		Morphological characteristics	Shannon
			Margalef
			Berger-Parker
		Pedigree	COD
			WCOD
	Country	Named varieties	Shannon
			Margalef
			Berger-Parker
			Number of varieties
		Morphological characteristics	Shannon
			Margalef
			Berger-Parker
		Pedigree	COD
			WCOD

Results from the calculations of all these diversity concepts are interesting to compare, but to lend an economic interpretation to the results, an understanding of the environment in which the changes took place is crucial. What factors caused the big swings and gradual changes? Diversity levels are not determined in isolation; instead, variety area shares represent the simultaneous solution of supply and demand for different varieties. They are the result of a combination of factors, including agro-ecological variation, policy decisions, reactions to policy decisions on the part of farmers and scientists, developments in research, and input and output prices. By exploring more deeply past influences on diversity levels, we will develop a better understanding of what is likely to affect them in the future. Also, by clarifying the definitions of diversity and the ways to measure them, we can more effectively investigate the relationship of diversity to risk and changes in the level of productivity.

REFERENCES

Benin, S., M. Smale, and J. Pender. 2006. Explaining the diversity of cereal crops and varieties grown on household farms in the highlands of northern Ethiopia. In Smale, M. (ed.). *Valuing Crop Diversity: On-farm Genetic Resources and Economic Change.* Wallingford: CABI Publishing.

Brennan, J.P. and D. Byerlee. 1991. The rate of crop varietal replacement on farms: Measures and empirical results for wheat. *Plant Varieties and Seeds* 4: 99-106.

Brush, S.B., J.E. Taylor, and M.R. Bellon. 1992. Technology adoption and biological diversity in Andean potato agriculture. *Journal of Development Economics* 39: 365-387.

Cox, T.S., J.P. Murphy, and D.M. Rodgers. 1986. Changes in genetic diversity in the red winter wheat regions of the United States. *Proceedings of the National Academy of Science* (USA) 83: 5583-5586.

Cox, T.S., Y.T. Kiang, M.B. Gorman, and D.M. Rodgers. 1985. Relationship between coefficient of parentage and genetic similarity indices in the soybean. *Crop Science* 25: 529-532.

Day Rubenstein, K., P. Heisey, R. Shoemaker, J. Sullivan, and G. Frisvold. 2005. *Crop Genetic Resources: An Economic Appraisal*. USDA/Economic Research Service: Economic Information Bulletin, Number 2.

Donini, P., J.R. Law, R.M.D. Koebner, and J.C. Reeves. 2000. Temporal trends in the diversity of UK wheat. *Theoretical Applied Genetics* 100: 912-917.

Dreisigacker, S., P. Zhang, M. Warburton, B. Skovmand, D. Hoisington, and A.E. Melchinger. 2004. SSR and pedigree analyses of genetic diversity among CIMMYT wheat lines targeted to different megaenvironments. *Crop Science* 44: 381-288.

Dudley, J. 1994. Comparison of genetic distance estimators using molecular marker data. In American Society for Horticultural Science (ASHS) and Crop Science Society of America (CSSA) (eds.), *Analysis of Molecular Marker Data, Joint Plant Breeding Symposia Series.* Corvallis, ASHS, CSSA.

Duvick, D.N. 1984. Genetic diversity in major farm crops on the farm and in reserve. *Economic Botany* 38: 161-178.

Edmeades, S., M. Smale, and D. Karamura. 2006. Demand for cultivar attributes and the biodiversity of bananas on farms in Uganda. In Smale, M. (ed.). *Valuing Crop Diversity: On-farm Genetic Resources and Economic Change.* Wallingford: CABI Publishing.

Franco, J., J. Crossa, J. Villaseñor, S. Taba, and S.A. Eberhart. 1998. Classifying genetic resources by categorical and continuous variables. *Crop Science* 38: 1688-1696.

Gauchan, D., M. Smale, N. Maxted, and M. Cole. 2006. Managing rice biodiversity on farms: The choices of farmers and breeders in Nepal. In Smale, M. (ed.). *Valuing Crop Diversity: On-farm Genetic Resources and Economic Change.* Wallingford: CABI Publishing.

Gebremedhin, B., M. Smale, and J. Pender. 2006. Determinants of cereal diversity in villages of northern Ethiopia. In Smale, M. (ed.). *Valuing Crop Diversity: On-farm Genetic Resources and Economic Change.* Wallingford: CABI Publishing.

Harper, J.L. and D.L. Hawksworth. 1995. Preface. In D.L. Hawksworth (ed.). *Biodiversity Measurement and Estimation.* The Royal Society and Chapman and Hall, London.

Hawksworth, D.L. (ed.). 1995. *Biodiversity Measurement and Estimation.* The Royal Society and Chapman and Hall, London.

Jin, S., J. Huang, R. Hu, and S. Rozelle. 2002. The creation and spread of technology and total factor productivity in China's agriculture. *American Journal of Agricultural Economics* 84: 916-930.

Kempthorne, O. 1969. *An Introduction to Genetic Statistics.* Ames: Iowa State University.

Koebner, R.M.D., P. Donini, J. Reeves, R.J. Cooke and J.R. Law. 2002. Temporal flux in the morphological and molecular diversity of UK barley. *Theoretical Applied Genetics* 106: 550-558.

Lage, J., M.L. Warburton, J. Crossa, B. Skovmand, and S.B. Anderson. 2003. Assessment of genetic diversity in synthetic hexaploid wheats and their *Triticum dicoccum* and *Aegilops tauschii* parents using AFLPs and agronomic traits. *Euphytica* 134: 305-317.

Ludwig, J.A. and J.F., Reynolds 1988. *Statistical Ecology: A Primer on Methods and Computing.* Wiley: New York.

Magurran, A. 1988. *Ecological Diversity and Its Measurement.* Princeton: Princeton University Press.

Malecot, G. 1948. *Les Mathematiques de l'Heredite.* Paris: Masson.

Meng, E.C.H., M. Smale, M.R. Bellon, and D. Grimanelli. 1998. Chapter 2: *Definition and Measurement of Crop Diversity for Economic Analysis.* Dordrecht and Mexico, D.F.: Kluwer Academic Publishers and CIMMYT.

Meng, E., M. Smale, H. Ruifa, J.P. Brennan, and D. Godden. 1999. Measuring crop genetic diversity for economic analysis: Concepts, indices and applications. Contributed paper, 43rd Annual Conference of the Australian Agricultural and Resource Economic Society, Christchurch, New Zealand.

Meng, E.C.H., 1997. *Land Allocation Decisions and in situ Conservation of Crop Genetic Resources: The Case of Wheat Landraces in Turkey.* Ph.D. thesis, University of California, Davis, California.

Reif, J.C., P. Zhang, S. Dreisigacker, M.L. Warburton, M. van Ginkel, D. Hoisington, M. Bohn, and A.E. Melchinger. 2005. Wheat genetic diversity trends during domestication and breeding. *Theoretical and Applied Genetics* 110: 859-864.

Smale, M. (ed.). 2006. *Valuing Crop Diversity: On-farm Genetic Resources and Economic Change.* Wallingford, UK: CABI Publishing.

Smale, M., M.R. Bellon, and J.A. Aguirre Gomez. 2001. Maize diversity, variety attributes, and farmers' choices in Southeastern Guanajuato. *Economic Development and Cultural Change* 50: 201-225.

Smale, M., E. Meng, J.P. Brennan, and R. Hu. 2003. Determinants of spatial diversity in modern wheat: Examples from Australia and China. *Agricultural Economics* 28: 13-26.

Smale, M., M.P. Reynolds, M. Warburton, B. Skovmand, R. Trethowan, R.P. Singh, I. Ortiz-Monasterio, and J. Crossa. 2002. Dimensions of Diversity in Modern Spring Bread Wheat in Developing Countries from 1965. *Crop Science* 42: 1766-1779.

Solow, A. and S. Polasky. 1994. *Measuring Biological Diversity.* Working Paper, Woods Hole Oceanographic Institution. Woods Hole: Woods Hole Oceanographic Institution.

Souza, E., P.N. Fox, D. Byerlee, and B. Skovmand. 1994. Spring wheat diversity in irrigated areas of two developing countries. *Crop Science* 34: 774-783.

Van Dusen, M.E. and J.E. Taylor. 2005. Missing Markets and Crop Diversity: Evidence from Mexico. *Environment and Development Economics* 10: 513-531.

Warburton, M.L., J. Crossa, J. Franco, M. Kazi, R. Trethowan, S. Rajaram, W. Pfeiffer, P. Zhang, S. Dreisigacker, and M. van Ginkel. 2006. Bringing wild relatives back into the family: recovering genetic diversity in CIMMYT improved wheat germplasm. *Euphytica* 149: 289-301.

Weitzman, M.L. 1992. On Diversity. *Quarterly Journal of Economics* 107: 363-405.

Zanatta, A.C.A., M. Keser, N. Kilinc. 1996. Agronomic performance of wheat landraces from western Turkey: Bases for in situ conservation practices by farmers. Paper presented at the 5th International Wheat Conference, Ankara, Turkey.

Chapter 3

Agricultural Policy, the Wheat Economy, and Crop Diversity in China

J. Huang and S. Rozelle

ABSTRACT

Government policies play a major role in crop diversity at the farm and aggregate levels in China. An examination of recent trends in the wheat economy in China reveals several significant policy changes that have altered outcomes in China's wheat economy: decollectivization, market reform, and associated changes in incentives. Those impacts have affected both the quantity of production and the wheat diversity observed. This chapter examines various elements of policy change and their impacts on wheat diversity. Changes in the research and extension systems and the household responsibility system are found to be among the most important influences on wheat crop diversity in China.

While the cultivation of a limited set of modern wheat varieties over an increasingly large area has raised concerns of genetic uniformity, many factors may have significant counterbalancing effects, including policies that influence agricultural production, consumption, and marketing. Policies in any of these areas may have direct and indirect influences on both the supply and demand of crop diversity. The growing literature of empirically-based research testing the relationship between agricultural policies and crop diversity suggests an influential role for decision makers at all levels of analysis.

A necessary condition for analyzing the influence of policies on crop diversity is an understanding of factors influencing agricultural productivity and growth. Economic reform and other changes in technology and institutions have caused rapid growth in production and incomes in many parts of China's rural sector since the late 1970s. Economists have focused on understanding the implications of the reforms and of technological change on agricultural growth during the period of reform (Lin 1992; Fan 1991; Huang and Rozelle 1996). The empirical record clearly demonstrates that institutional reform—in particular, decollectivization and associated changes in property rights—contributed significantly to the growth of agriculture in the early reform period. After the mid-1980s, technological change took over as the dominant source of agricultural growth. Studies on technology-production linkages suggest that plant breeding and other varietal improvements have been significant factors in agricultural growth (Hu et al. 2000; Jin et al. 2002; Huang and Rozelle 1996).

In addition to these traditional sources of growth, the contributions of crop diversity to crop productivity and production variability have also been explored (see Chapters 10 and 11; Hu et al. 2000). Several empirical studies on the genetic diversity of major food crops (rice, wheat, maize, and soybean) in China suggest that the widespread cultivation of modern varieties has not resulted in increasing uniformity over time, according to various concepts and taxonomies of diversity (Chapter 6; Wang et al. 2001; Hu et al. 2000). Using named varieties as the basis of analysis, the pool of major wheat varieties cultivated by farmers in 14 of China's major wheat provinces increased from 211 in 1982 to well over 300 in the 1990s (Jin et al. 2002). Pedigree-based measures as well as those based on morphological characteristics confirm the increasing trend in wheat diversity at a national level over the same time period (Chapter 6).

Progress has also been made in identifying the determinants of spatial diversity at regional levels (Chapter 8). The complexity of the determinants of crop diversity requires the application of multivariate analyses to better understand how policies affect cultivar distribution across provinces. Analysis of expected household impacts is also needed. Many factors affect diversity at the household level, but we believe that policies are particularly important.

The overall goal of this chapter is to identify national agricultural and agriculturally-related policies likely to affect the demand and supply of crop diversity in China and to examine the direction of the impact. Policies reviewed in this chapter include those with direct impacts on agricultural production and those that have indirect impacts on agricultural production through the demand for and marketing of agricultural products. The focus of this study is wheat, but the discussions will extend to policies likely to have similar impacts on a range of crops. In this chapter, we first discuss recent trends in the wheat economy of China to provide a general overview of China's wheat sector that covers both production and consumption issues. Next, we describe the major policies that govern China's wheat economy. The impact of these policies on wheat production is examined, and finally, the implications for wheat diversity are hypothesized.

CHINA'S WHEAT ECONOMY

Production

Wheat production in China has expanded steadily throughout the past several decades. Even between 1970 and 1978, when the Cultural Revolution stifled economic activity in other parts of the economy, wheat production grew at 7% per annum (Table 3.1). However, after accelerating to 8.3% per year in the early reform period (1978-84), wheat production growth slowed significantly during the 1990s and has been negative during 1998-2003. Growth rates in yields have been largely similar to production growth rates. In the 1970s and the early reform period, wheat yields increased at annual rates exceeding those of rice and maize. Wheat producers also maintained their sown area, whereas rice producers' average sown area fell 0.6% per year between 1978 and 1995. More recently, wheat production area has fallen and maize area increased over every period preceding 1999-2003.

Table 3.1 Annual growth rates of production, sown area, and yields of wheat, rice, and maize in China, 1970-2003.

Commodity	*Pre-reform*	*Reform period*			
	1970-78	1978-84	1985-90	1991-98	1999-2003
	% p.a.	*% p.a.*	*% p.a.*	*% p.a.*	*% p.a.*
Wheat					
Production	7.0	8.3	1.4	2.2	–5.4
Sown area	1.7	–0.0	0.5	–0.5	–6.0
Yield	5.2	8.3	0.9	2.8	0.6
Rice					
Production	2.5	4.5	1.1	1.0	–4.2
Sown area	0.7	–0.6	0.1	–0.6	–3.3
Yield	1.8	5.1	1.0	1.6	–1.0
Maize					
Production	7.4	3.7	4.1	4.1	–2.2
Sown area	3.1	–1.6	2.7	2.3	–1.0
Yield	4.2	5.4	1.4	1.7	–1.3

Note: Growth rates are computed using regression method.
Source: NSBC 1980-2002.

Farmers grow wheat in every province of China, but cropping patterns that include wheat, as well as the intensity and importance of wheat, vary from region to region. With the exception of single-season spring wheat in the four northern-most provinces (Heilongjiang, Jilin, Liaoning, and Inner Mongolia), farmers produce wheat in tight rotations with other crops. Farmers in the North China Plain most commonly plant winter wheat in conjunction with maize or cotton. Because timely planting of maize and other crops is necessary to avoid fall frosts, wheat farmers traditionally have left space between wheat rows to enable sowing or transplanting of maize or cotton crops prior to the wheat harvest. Yangtze Valley farmers, especially those living north of the river in regions where two rice crops do poorly, use facultative, over-wintering wheat in rotation with single-season rice crops.

China's wheat basket is in the northern maize-wheat region, which includes the three officially defined geographic regions of North, Northeast, and Northwest China. In 1975, 68% of wheat was sown in the North China Plain, the Northeast region, and several Northwest provinces. Although still dominant, the proportion of wheat sown in North China declined somewhat in the 1990s, primarily because many farmers in the more southern Yangtze Valley region moved from two-season rice to rice-wheat rotations. Nevertheless, all but three of the provinces that produce the most wheat are located in northern China.

Demand

Wheat is used in many ways in China. Direct food consumption accounted for 81% of use in the 1990s. Seed use, post-harvest waste, and feed and industry demand together accounted for the remainder (Table 3.2). In urban areas, per capita wheat consumption has been declining, as meat and non-staple food consumption have risen since the early 1980s. Rural residents continued to expand their wheat consumption during the 1980s and most of

Table 3.2 Annual wheat supply and utilization food balance sheet in China, 1981-2000.

	Units	1981-85	1986-90	1991-95	1996-2000
Crop area	'000 ha	28,821	29,559	29,904	28,990
Yield	t/ha	2.66	3.06	3.38	3.84
Production	'000 t	76,625	90,455	101,087	111,420
Stock change	'000 t	492	–512	–320	636
Net import	'000 t	10,716	12,434	9,342	2,043
Import	'000 t	10,769	12,517	9,737	2,585
Export	'000 t	53	84	395	542
Consumption	'000 t	86,849	103,401	110,749	112,828
Food use	%	78	80	81	81
Feed use	%	3	3	3	3
Seed use	%	10	8	7	6
Industry use	%	3	3	3	3
Waste	%	6	6	6	6
Per capita food	kg/person	66.2	75.0	75.9	73.6
Urban	kg/person	60.6	59.0	55.2	45.7
Rural	kg/person	67.9	80.4	83.9	85.7
Self-sufficient level	%	88	87	91	99

Source: CCAP database and authors' estimates.

the 1990s, although per capita wheat consumption seems to have reached a peak by the late 1990s. The average annual per capita wheat consumption in rural areas was only about two kilograms higher during 1996-2000 than during 1991-95 (Table 3.2). The impact of rural-urban migration on wheat demand is likely to differ in China from that in other Asian countries. China is a major wheat producer with a rural economy that consumes much more wheat than its urban economy, particularly in the northern regions where wheat is the main staple food. Rural to urban migration is expected to reduce the national demand for wheat and, in fact, has done so since the mid-1990s (Table 3.2).

International Trade

China has been one of the world's largest wheat importers. Annual figures for wheat imports from the 1980s to the mid-1990s ranged from 5 to 15 million tons. Wheat imports during this period have exhibited a cyclical pattern, whereby an increase in imports every three or four years is followed by a decrease lasting for three or four years. This pattern of trade is linked closely both with domestic wheat production and with trade policies. However, the longer-term trend for annual imports has been a gradual decline from nearly 13 million tons in 1986-90 to less than 3 million tons in 1996-2000 (Table 3.2). A major factor in this long-term decline in imports has been a slowing in food consumption growth coinciding with periods of stable wheat production.

Government Interventions in China's Agriculture

China is a rapidly developing country in transition from a socialist system to one in which an increasing proportion of its goods and services, including food, are being allocated by prices and other market forces (Sicular 1995). China's government, however, remains deeply involved in guiding the nation's development. Food security has been and will continue to be the central goal of China's agricultural policy. The Tenth Five-year Plan for 2001-05 and the National Long-Term Economic Plan for the next 15 years both call for a number of different goals: continued agricultural production; rising farmer incomes; the maintenance of near self-sufficiency for food grains (rice and wheat); and the elimination of absolute poverty. These policies along with many forces arising from development and transition will fundamentally shape China's wheat supply and demand, as well as carrying implications for wheat diversity at the farm and aggregate levels.

Institutional Reform

China first implemented decollectivization policies in the late 1970s, focusing first on poorer regions and gradually extending the policies to the whole country. By 1980, 14% of China's villages had returned land use rights to farm households (the Household Responsibility System), a figure that moved rapidly upward in the early 1980s, reaching and remaining at 99% of villages in 1985 (Table 3.3, last column). After decollectivization, every farm household in China was endowed with land. Although land ownership officially rests with the village, local leaders contract the land (without compensation for the most part) to village households. Legal tenure security on contracted land was extended from 15 to 30 years in the late 1990s. In practice, however, during the past two decades the dynamics of household and village demographics and other policy pressures have often induced local authorities to reallocate land prior to the contract expiration period. The Rural Land Contract Law, new legislation that took effect in March 2003, is intended to address issues of land insecurity as well as clarifying rights for the transfer, exchange, and even inheritance of contracted land. These measures are intended to improve land use efficiency, at least partially through increases in farm size.

Providing land to all farmers has essentially given an income-producing asset to every household, with associated positive effects on food security and reduced poverty. While there is probably no single policy more responsible for eliminating China's poverty, land fragmentation and the small size of farms constrain the growth of labor productivity and farmer income. Probably more than any other single feature, the size of farms in China defines its agriculture. In 1980, the average size was only 0.56 hectare per farm (approximately 0.15 hectare per capita) split into 4-5 plots (often with varying quality) located in different reaches of the village. By 1999, farm size fell to 0.40 hectare (Huang 2000). Despite their small size, China's farms still produce more than half the income of the average rural household. Each farm household has an income from farming, but the rise in non-agricultural income has contributed most of the gains in per capita rural incomes during the reform era. Research shows that work off-farm presents one of the most likely means for rural residents to escape poverty (World Bank 2001).

Technology: Research, Extension, and Seed Reform

Technological change has been the engine of China's agricultural economy in general, and for small grains like rice and wheat in particular (Stone 1988). While less dramatic than the well-known story of the discovery and extension of hybrid rice (Lin 1992), continuous and rapid technical change came to wheat farmers in recent decades. After importing rust resistant, semi-dwarf varieties from the international agricultural research system in the late 1960s, China's breeders incorporated the traits into their own varieties. As of 1977, producers were sowing about 40% of China's wheat area to semi-dwarf varieties; by 1984, this number had risen to 70% (Rozelle and Huang 2000). Currently it is difficult to find anything but improved varieties in China's wheat area. Certainly the rapid diffusion of new technology contributed to wheat yield growth in the reform era.

Table 3.3 Important factors affecting wheat supply in China, selected years, 1975-95.

Year	*Labor*		*Agricultural research*[1] *(million yuan)*		*Irrigation (million yuan)*		*Price (yuan/t)*		*Household Responsibility System ratio*
	Input (days/ha)	Wage (yuan/day)	Expenditure	Stock	Expenditure	Stock	Wheat market	Fertilizer mixed	
1975	402	2.60	2,004	1,006	9,690	97,455	1,397	842	0.00
1980	347	2.67	2,229	1,445	9,040	134,727	1,859	792	0.14
1985	222	4.23	4,044	2,030	5,679	140,668	1,316	1,042	0.99
1990	210	5.04	3,774	2,798	8,474	150,669	1,603	1,042	0.99
1995	190	7.39	5,214	3,478	17,743	200,967	1,713	1,978	0.99
1999	157	10.55	6,768	3,856	29,684	268,364	1,431	1,730	1.00

Note: Values are in 1999 prices.

Sources: Irrigation data computed by authors based from MWR (1988-2000). Research expenditure data are from SSTC. Prices are from State Price Bureau. Labor input and wage are computed by authors based on data from SPB (1978-2000).

[1] *The research stock variable was estimated as* $Z_R(t) = \Sigma_{\tau=0}^{n} \alpha(t)\, E_R(t)$, *where* $E_R(t)$ *is the research expenditure in period t and* $\alpha(t)$ *is a vector of weights lagged for 17 years for timing the accumulation of new research expenditures to the stock of research. Because most of the formal agricultural research programs were started in the late 1960s, the research stock generated from these investments was small and only 1006 million yuan. Given the vector weights, it is expected that on average, research stock in year t would approach the annual expenditure in approximately year t-15.*

Robust growth in the stock of research capital has in part been responsible for these dramatic changes (Fan and Pardey 1995; Huang et al. 1996). Once the model of developing country research systems, China's agricultural research programs, however, may be suffering from neglect after more than a decade of its own reform (Rozelle et al. 1997). Real annual expenditures on agricultural research fell between 1985 and 1990 before resuming real growth in 1990 (Table 3.3, column 4). The slowdown in growth in annual investments in the 1980s resulted in slower growth in the overall stock of research in the 1990s (column 5). Jin (1997), however, shows that if economic indicators signal tightening supplies and rising prices, officials will respond by increasing current expenditures. Public investment in agricultural research increased in the late 1990s due to growing concern about China's ability to meet future demand for agricultural products. Within the research system, policy changes have occurred in three systems: agricultural research, extension, and seed production/supply.

Research system reform

As part of China's general move to distance itself from the planning system, reformers gradually implemented a series of science and technology policies designed to alter the behavior and output of research institutes. In addition to opening to the outside world, the agricultural research reforms of the 1980s and early 1990s targeted two main areas: (1) changes in the distribution of research funds to a more competitive system, focusing resources on the most productive scholars and institutes; and (2) policies encouraging research institutes to commercialize their products, allowing them to retain profits and reinvest as a major source of revenue. Currently, most research funds from national sources can be accessed only through competitive research funding programs for priority research areas. Government objectives in areas of social and environmental concern (food security, poverty alleviation, and environmental protection) can also be easily incorporated into competitive research programs. Assessment of the early stage of agricultural research reforms found that they were only partially successful (Rozelle et al. 1996). Despite the shortcomings, in the mid-to-late 1990s there were signs that the reforms were beginning to move the research system toward a more market-oriented model.

Dissatisfaction with the perceived benefits of earlier reforms in terms of providing new technologies to producers, duplication of research among institutes, and the continued existence of an over-staffed and under-funded research system created a new impetus to launch another round of reforms. In addition, the shifting needs created by China's move to a more market-oriented economy and the challenges of research in the new high technology fields reinforced the need to reform the agricultural research system. In the new round, officials have set high goals: the creation of a modern, responsive, internationally competitive and financially sustainable agricultural research innovation system. The new round of reforms attempts to divide activities of current staff into those with potential commercial applications and those that constitute more applied-basic and basic research. Scientists in the non-commercial sector are categorized according to research ability and potential, and those that are prioritized are provided with higher salaries, a large increase in per capita support, and new facilities and equipment.

Extension

The extension system has been facing an even more serious crisis than the agricultural research system. With millions of extension agents in a system that is under-funded, leaders have turned to quasi-commercialization to allow the system to survive. While the extension system is still charged with the dissemination of new technology to farmers, salaries for extension agents have either been eliminated or drastically cut back. In return, extension agents are allowed to do business in fertilizer, pesticide, and seed sales to supplement their income. The results of these policies, however, have not been productive. According to most observers, services have been reduced, and in many cases the fact that agents simultaneously represent an input supply company has led to their providing counter-productive advice.

Seeds

Efforts to build a national seed system began in the 1950s, and China's seed production and distribution system is now the largest in the world (Hu 1995). The state seed supply organization is now partially commercialized and in the beginning of the 21st century consisted of approximately 2,200 county seed companies, 500 prefectural seed companies, 30 provincial seed companies, the National China Seed Corporation, as well as hundreds of seed companies owned by the public plant breeding institute and other agricultural research organizations and universities.

Rules and regulatory institutions to administer the seed industry have evolved within the Ministry of Agriculture over the past several decades (World Bank 1996). With the stated objective of ensuring seed quality, ownership of seed companies that produced and distributed seed of hybrid rice and maize was limited to state-owned enterprises until the late 1990s. As part of a reform package to address in part the financial stress affecting many public agricultural research units, China allowed research institutes and universities to distribute the hybrid varieties produced in their own breeding programs. Private firms, initially not included in this reform package, were later given permission to do the same. These reforms, however, applied exclusively to materials developed internally.

Meaningful reform of the seed industry and related legislation did not begin until after the mid-1990s, later than in almost any other subsector. Recent changes in the laws that govern the seed industry are now promoting the commercialization of the seed industry in general by encouraging the entry of new domestic firms. Foreign investment in the seed industry has slowly been permitted. In 1997, China also passed a Plant Variety Protection Act and signed the UPOV agreement. A few large Chinese firms have been allowed to raise money by selling some of their shares on the stock market. The seed law implemented in 2000 defined the role of the private sector and cleared the way for any investor meeting the minimum requirement for capital investment and facilities to sell seed. Companies are now allowed to sell seed of hybrid maize, rice, and cotton varieties bred by public institutes, thus eroding local monopolies long held by county seed companies. The burden of bureaucratic requirements has also decreased: companies meeting certain requirements can now obtain permits to sell seed in all counties of a province or in all provinces in the country, instead of being forced to apply separately at each administrative level.

Despite the significant reforms, numerous constraints to the development of China's seed industry still continue. Thousands of small, local seed companies, many of them publicly owned, still dominate the industry. Although markets can be competitive within a region, local markets are often artificially isolated by measures adopted by local governments so that only small local firms are able to participate. Seed quality and related services provided also vary across regions. The obstruction of farmer access to a larger selection of new varieties negatively impacts efficiency, and limits on the market size of larger firms may affect incentives for research and development. As a consequence, the system is likely to result in the slow spread of major new varieties across large regions.

The current seed system also appears to affect the rights and ability of breeders to profit from the development and sale of their varieties. Seed regulations require that the breeder submit parent lines and all breeding information at the start of the registration process. With the information in the public domain, and the foundation seed in the hands of seed companies, breeding institutes earn limited revenue from their varieties over the long run, a factor that has reduced the incentives of breeders to innovate. The lack of separation of policy functions and commercial activities is a significant problem facing seed industry managers who, under complete liberalization, might otherwise improve the efficiency and service orientation of their firms (Rozelle et al. 1997).

Production Environment

Irrigation investment

China's progress in water control also has contributed to the cropping sector's productivity gains (Liu 1992). Irrigated area increased from less than 18% of cultivated area in 1952 to more than 50% in the later 1990s (MWR). In the initial years, most of the construction was based both on locally organized small-scale projects and publicly financed large-scale surface projects (Stone

1993). In the late 1960s and 1970s, tubewell development drove the expansion of irrigated area construction, especially in the North China Plain maize-wheat region (Wang et al. 2000). Development of the nation's water control infrastructure has continued during the 1980s with a large number of government-sponsored medium- and larger-scale water control projects. Even though pump numbers stagnated in the 1980s, the overall quality of water control equipment has continually improved (MWR). Irrigation has been one of the major factors influencing land and labor utilization in the cropping sector in the 1970s and 1980s, as improvements in water control stimulated the increase in double-cropped areas.

Environmental degradation

Environmental degradation, particularly wind and water erosion and salinization, has increased significantly in China over the past several decades. Estimates of the land affected by water and wind erosion range from 20% to 30% (Peng and Xu 1993). Studies have shown a reduction of as much as 25% in grain yields and other agricultural outputs during 1976-95 (Huang and Rozelle 1995, 1996; Rozelle et al. 1997).

Food Price and Marketing Policies

Price and market reforms are key components of China's development policy shift from a socialist to a market-oriented economy. The reforms have been implemented gradually with liberalization of non-strategic commodities, such as vegetables, fruits, oilseeds, sugar, livestock, and fish, taking place in the early reform years. Only later in the reform process (during the 1990s) were more strategic commodities such as rice, wheat, maize, and cotton affected. However, while the reforms for non-strategic commodities were implemented smoothly and successfully in the 1980s, the reforms in the grain sector have taken place in a stop-start pattern. After a record growth in agricultural production in 1984 and 1985, price and marketing reforms were announced in 1985 to limit radically the scope of government price and market interventions and further enlarge the role of market allocation. Because of the sharp drop in the growth of agricultural production and grain price inflation in the late 1980s, however, implementation of marketing reform stalled. Mandatory procurement of wheat, rice, maize, oil crops, and cotton continued. To provide more incentive for farmers to raise productivity and sell to the government, contract prices for wheat and other grains were raised over time. Despite this, the increases in the nominal wheat and other grain procurement prices were lower than the inflation rate and led to a decline in real farm gate grain prices.

As grain production and prices stabilized in the early 1990s, another attempt was made in early 1993 to abolish the compulsory grain quota system and the sale of grain to consumers at below-market prices. While both the state grain distribution and procurement systems were substantially liberalized, the policy was reversed when food price inflation reappeared in 1994. Government grain procurement once again became compulsory. The provincial governor's "Rice Bag" responsibility system that emphasized local self-sufficiency in rice, wheat, and maize was introduced over 1994-95. With the slowdown in the growth of farmer incomes and the increasing fiscal burden of financing grain marketing after the mid-1990s, the central government initiated a controversial policy in the grain marketing system in 1998. Under the 1998 policy, individuals and private companies were prohibited from procuring grain from farmers. The ban on private grain procurement was considered by the government as a pre-condition to eliminating the government's financial burden. Grain quota procurement prices were set at higher-than-market prices. Prices of grain sold by grain bureaus directly to markets or to private traders were supposed to be set at a level higher than procurement prices, to cover marketing operation costs and therefore to avoid losses in marketing by grain bureaus.

None of the original policy goals was achieved after three years of implementing this policy, as demonstrated by the amount of free market grain purchases in Table 3.4. The high costs of monitoring and inspecting grain markets and the unwillingness of the grain bureaus to pay the protection prices to farmers meant that private traders continued to purchase grain from farmers, even as the policy was being implemented (Huang 2001). Largely as a result of the government's increasing fiscal burden, the grain procurement quota was first eliminated in grain deficit areas in 2000 and in other regions beginning in 2001, with an ultimate objective of completely phasing out the government procurement program.

The grain procurement policy in general has taxed grain farmers heavily, although marketing reforms and the development of grain markets in China have begun to ease this burden (Huang et al. 2004). Developments in 2004, however, signal a significant policy change. Grain subsidies in the form of direct payments, the elimination of agricultural taxes and taxes on specialty crops, and seed subsidies for high quality grain are all included as agricultural policy instruments to raise farmer incomes and ensure grain production (Gale et al. 2005).

Foreign Exchange and Trade Policies

China's open door policy contributed to the rapid growth of the external economy and to a greater reliance on both domestic and international trade to meet consumer demand. Historically, the overvaluation of the domestic currency for trade protection purposes reduced agricultural incentives. Real exchange rates remained constant and even appreciated during the 30 years prior to reforms. After reform, however, the exchange rate depreciated rapidly in every year, with the exception of several years of the domestic price inflation

Table 3.4 Government grain procurement and market development in China, 1979-98.

Year	Grain production (m.t)	Total marketed grain		Urban population %	Gov't procurement (m.t.)			Free market (m.t)	Market share (%)		
		Amount (m.t)	% of prod'n		Total	Quota	Negotiated		Gov't quota	Gov't negotiated	Free market
	(1)	*(2)*	*(3)*	*(4)*	*(5)*	*(6)*	*(7)*	*(8)*	*(9)*	*(10)*	*(11)*
1979	289	64	22	19	59	54	5	5	84	8	7
1980	279	64	23	19	59	50	9	5	78	13	8
1981	282	68	24	20	63	52	11	6	76	16	8
1982	306	80	26	21	74	56	18	6	71	22	7
1983	337	105	31	22	99	91	8	6	87	7	6
1984	354	118	33	23	111	102	9	7	86	8	6
1985	329	85	26	24	79	60	20	6	70	23	7
1986	340	102	30	25	95	62	32	8	61	32	8
1987	351	108	31	25	99	57	42	9	53	39	8
1988	343	104	30	26	94	50	45	10	47	43	10
1989	354	111	31	26	101	49	52	11	44	46	10
1990	389	122	31	26	96	52	44	27	42	36	22
1991	380	120	31	27	99	47	52	21	39	43	18
1992	387	121	31	28	97	45	52	25	37	42	21
1993	403	131	33	28	90	51	39	41	39	30	31
1994	392	128	33	29	89	44	45	38	35	35	30
1995	411	140	34	29	93	46	46	47	33	33	34
1996	446	165	37	29	98	50	47	67	30	29	41
1997	434	165	38	30	110	50	60	55	30	36	33
1998	453	145	32	30	96	50	46	49	35	32	34

Note: Quota procurement included both quota and above-quota (with higher price than quota) procurements before 1985 and contract procurement thereafter.

Sources: Zhongguo Gongshang Xingzheng Guanli Tongji Sishinian [Forty Years of Statistics of China's Market Administration and Management]. State Statistical Bureau Press: Beijing, China. 1992; and *Zhongguo Guonei Maoyibu Nianjian* [China Ministry of Domestic Trade Yearbook]. State Statistical Bureau Press: Beijing, China. 1995-1999. Data for 1997 and 1998 are based on the authors' interviews.

during the mid-1980s. Falling exchange rates increased export competitiveness and contributed to China's phenomenal export growth record (i.e., in non-grain food products) and its impressive national economic performance of the 1980s. China's wheat price protections were also reduced in this period due to falling exchange rates, though impacts on wheat trade were somewhat diminished by the monopoly of the government's state trading agent. In the 1990s, however, the real exchange rate appreciated by about 25% (IMF 2002).

Wheat is a much less homogeneous commodity than other traded commodities such as maize and soybean. During a wheat market survey conducted in 2001, traders reported that the price of high quality wheat from North America was 20% to 50% higher in the domestic markets of China's major ports than its price upon arriving at the port. However, though growing, the market for baking-quality wheat is still relatively small in China. Traders reported the CIF price of medium quality wheat (used for more common bread, cheaper pastries, industrial uses, and high quality noodles) imports from Australia, England, and the Pacific Northwest of the United States to be 10% lower than the price that they believed the same wheat would command in China's domestic market. Traders interviewed estimated that this market accounted for around 10% to 15% of China's wheat demand. Although there have been no imports of the low (or lower-medium) quality wheat that make up the biggest part of China's wheat production (estimated to be more than 60%), this segment of the market is only marginally protected (Huang et al. 2004). The price of China's lowest quality of wheat (about 10% to 15% of its harvest) was largely the same as the world's feed wheat price. In fact, from 2001 to 2003, China exported feed wheat into international markets.

IMPACTS OF AGRICULTURAL POLICIES ON WHEAT PRODUCTION AND GENETIC DIVERSITY

Impacts of Policies and Sources of Wheat Production Growth

In an economy such as China's, wheat output is affected by multiple factors. Rozelle and Huang (2000) show that investments in agricultural research and extension have been two of the most important determinants of growth in wheat production. While irrigation availability positively affects wheat production in North China where water is a major constraint of crop production, increased irrigation does not raise wheat production in the regions without water shortage problems. Our earlier study also found that institutional reform such as decollectivization had a positive impact on the production of all crops, including wheat, reinforcing results found by other studies (Lin 1992; McMillan et al. 1989; Fan 1991; Huang and Rozelle 1996). Decollectivization-led output increases, however, were not a result of increased labor use. Consistent with the labor use data in Table 3.3, the institutional reforms led to substantially lower labor use. To compensate, wheat farmers in the post-reform period used chemical fertilizers to substitute for falling labor input and the increasing scarcity of land, a trade-off first described by Ye and Rozelle (1994) in their study of Jiangsu rice farmers.

Growth decomposition analysis shows that while institutional innovations are important, government investments have contributed the most to wheat yield growth during China's reform period (Rozelle and Huang 2000). Improvements in technology from research expenditures have by far contributed the largest share. Between 1976 and 1995 the implementation of the household responsibility system was the second-most important factor. The relatively high return to technology, however, had important implications for policy makers in China, who in the 1980s appeared to have believed that China could maintain its fast growth on the basis of institutional change, and consequently ignored research and water control investments. Rozelle and Huang (2000) also show that the net impact of fertilizer and wheat price changes is marginal for the whole reform period, as the increase in fertilizer prices cancels out the positive effect on wheat output of real increases in wheat prices. The positive impacts of government investment and institutional reform policies were also partially offset by increases in land and labor prices. However, given the massive shifts of labor out of wheat farming, the drop is only small, most likely due to the efficiency gains from the household responsibility system.

In terms of future trends, Rozelle and Huang (2000) forecast that production growth would keep up with the expected growth in demand. After reaching a peak in the mid-1990s (about 15 million tons), imports are projected to gradually decline into the early 21st century, mainly due to urbanization, declining population growth rates, and relatively low and falling income elasticities for wheat. As supply growth is sustained with the ongoing recovery of investment in agricultural research and irrigation, supply is projected to speed up and to slowly begin to meet most of national demand by 2020.

Impacts of Policy on Wheat Diversity

There is a rich literature on the impact of various policies on wheat production and consumption and their implications for the future demand, supply, and trade of wheat in China, but little information is available on the potential effects of these agricultural and agriculturally-related policies on wheat diversity. Our discussions here on the impacts of policies on wheat diversity are based on expectations and findings from the existing literature, and raise hypotheses that require further testing with available data. As a framework for the discussion, we categorize the policies outlined in the previous section into five groups: (1) institutional changes and fiscal policies, (2) technology policies, (3) irrigation expansion, (4) marketing reforms, and (5) trade liberalization.

Institutional changes and rural poverty reforms

The household responsibility system (HRS) reform was initiated in 1978 and resulted in the distribution of nearly all collectively-owned land to individual farmers. This change is likely to have had significant impacts on wheat diversity at both the farm and aggregate regional levels. Prior to the HRS, a uniform top-down approach was used to spread agricultural production technology. The technology extension station and county as well as commune (later renamed as township) governments were solely responsible for the adoption of crop varieties. Following an official decision on the variety or varieties to be planted in each production team and village, seed for the selected varieties was distributed by the agricultural extension station through the commune to village and production teams under a planned system. Collective production had a comparative advantage in the rapid expansion of new and high-yielding varieties, but also tended to offer fewer crop varieties at a given time, because the crop seeds were distributed through the "unified seed supply" of each agricultural extension station. Under this production system, the concept of household-maintained diversity did not exist and villages often used only one or two wheat varieties.

To the extent that it shifted the varietal adoption decisions to hundreds of millions of small individual households, decollectivization probably had a positive impact on wheat diversity relative to the pre-HRS period. Individual households are likely to have more heterogeneous wheat production and consumption demands than commune officials. Analysis from a farm-level survey in three major wheat producing provinces in China shows that households planted from one to four varieties with specific objectives and rationales for each choice (Chapter 9). Low-quality, high-yielding wheat varieties were often planted to fulfill the grain procurement quota, while higher-quality wheat varieties were cultivated for households' own food consumption needs. Households also planted additional varieties to meet various home-based food processing uses.

Decollectivization, however, may also have led to a lower level and/or slower rate of adoption of new varieties, due to a decrease in the ability of all farmers to equally access new technologies in the post-decollectivization period. The capability of the research system to deliver technology as widely and as uniformly to all villages and across all farmers in a village also declined. Lin (1992) and Huang and Rozelle (1996) find both positive and negative influences for adoption of hybrid rice after the reforms in the early 1980s. Research on household choice of wheat varieties also confirmed that the degree of production and marketing risks encountered at the household level induces a higher level of wheat diversity maintained by the household (Chapter 9). The existence of heterogeneous adoption decisions at the household level, as well as differing rates of household varietal turnover, contributes to heterogeneity in the aggregate pool of cultivated varieties. However, diversity outcomes at the household and aggregate levels will also depend on the diversity in the genetic materials used by breeding programs and on the success of related institutions in multiplying and marketing seed.

More recent reforms, such as the 2003 Rural Land Contract Law, could result in an increased level of uniformity in cultivated varieties by consolidating land that is currently spread across different households and effectively reducing the number of decision makers. The changing nature of the rural labor supply as a result of the complex interaction of other institutional changes may also have an indirect effect on diversity. As the opportunity cost of labor rises and more time is spent pursuing off-farm income, the time available for managing several different varieties within a household may decrease. The rising opportunity costs of labor negatively influences wheat diversity at the household level (Chapter 9).

Finally, the effects on farm income and agricultural production decisions of the recently implemented agricultural subsidies for grain production and the elimination of agricultural taxes remain certain. The new seed subsidization policy, on the other hand, could have a much more direct effect on wheat diversity, depending on the number of wheat varieties eligible for the subsidy and on the successful implementation of the policy.

Technology policies and reform

China's technology policies could potentially have both positive and negative effects on wheat diversity. Given the increasing profit and commercial orientation of the research system, there may be a greater incentive to focus research efforts on a more limited client base that better guarantees commercial profit. The outcome of this emphasis may be a system with little interest in creating a suitable range of wheat varieties with characteristics for niche markets or for marginal areas. Some parallels can be drawn with the trends in market deregulation and increasingly commercialized breeding in Australia (Chapter 4). Information from the breeders' survey in Australia regarding increasing commercialization and funding pressures indicates that a likely breeder response will be a focus on short-term outcomes at the possible expense of breeding strategies actively taking crop diversity into consideration (Chapter 5). Conversely, differentiating for these niche markets with specialty quality or other attributes will drive efforts to achieve commercial success and profit in the future. The shift of focus onto wheat quality as a research priority along with yield has likely resulted in the inclusion of new sources of genetic variation for the traits of interest (Chapter 6).

Similarly, in assessing the likely impact on wheat diversity of reforms in the seed industry, increasing commercialization and pressures for profit stand out. If, as a result of these pressures, investment and efforts focus increasingly on the limited pool of varieties with the highest expected profit, a greater concentration of varieties would be the outcome. However, even under that scenario, actual diversity at the household and aggregate levels will also depend on the variation in the genetic materials of that pool of varieties.

There is little empirical evidence on research reforms or the seed industry, but we have found mixed results from interactions between extension and diversity (Rozelle et al. 2003). In particular, in systems that have given greater support to extension, there tends to be a negative impact on diversity (in terms of diversity of both named varieties and of morphological traits), due to a focus on a relatively small set of preferred varieties. Within China, extension policies have been heavily influenced by broader grain policies that focused on a small number of priority commodities for food security and, within those commodities, emphasized yield increases. To the extent that extension reforms are captured by greater levels of funding, then our results suggest that, as the extension system strengthens, it may concentrate more on fewer varieties and reduce demand for diversity. If so, additional policies targeted at maintaining or increasing diversity may be needed to encourage the promotion of varieties for niche markets and for use in other areas that large-scale commercial concerns may find less attractive.

Irrigation expansion and environmental degradation

While increased irrigation positively affects productivity, the effects on diversity are more complex. Research in areas where traditional varieties of wheat are cultivated show that irrigation availability significantly influences plot-level decisions in favor of modern varieties (Rozelle et al. 2000). Varietal use decisions in Chinese households occur largely in a context dominated by modern varieties, so it would not be unreasonable to expect a significant impact of irrigation on variety choice. The expansion of irrigated area in China has also been significantly associated with a reduction in the use of marginal lands (Qiao 1997). Increased environmental homogeneity is likely to have a negative effect on overall diversity, as the need for local adaptation decreases and a smaller number of varieties becomes more suitable for larger areas. It is also likely that improved irrigation availability will increase the diversification into alternative crops as a means of reducing risk and/or increasing expected profit. On the other hand, it is possible that the wider availability of irrigated land could provide farmers with more flexibility in variety choices, as they retain more control over the cultivation environment. Farmers would also have access to a wider range of varieties, since the current supply-oriented research system has traditionally focused on producing varieties for irrigated areas.

Inasmuch as environmental heterogeneity has been positively associated with greater diversity, it is possible that the increasing degradation caused by erosion and salinization may have a positive effect on wheat diversity. Unless wheat varieties are purposely bred for adaptability to a wide range of environments, certain varieties may be more suitable to particular environmental conditions than others and selected based on expression of the necessary traits for those local conditions.

Food price and marketing reform

Rozelle et al. (2003) found mixed evidence of the effect of marketing reform and the emergence of markets on diversity. Since the mid-1980s, price and marketing reforms have been implemented gradually (de Brauw et al. 2004). The overall goal of the reforms is to create an effective domestic grain market by gradually phasing out the state grain quota procurement and distribution and by reducing price distortions at various market levels (from farm-gate to wholesale and retail). Implications of grain market reform for wheat diversity again are complicated, and overall impacts are ambiguous. Grain procurement quotas have promoted the adoption of low-quality, high-yielding varieties and discouraged the diversification of wheat varieties due to the lack of variability of prices among various qualities of wheat. However, the recent elimination of the grain procurement system and the increased numbers and activity of traders are likely to have a countering effect. Grain market reform and market infrastructure development that have improved grain grading and standard systems should also increase the market's ability to meet diversified demands for various wheat products. However, it is also possible that market liberalization and more commercial farming systems might induce farmers to concentrate on fewer varieties that are more profitable and marketable.

More commercially-oriented food market policies will affect demand indirectly by improving China's marketing environment and increasing the number of urban and rural markets. Changes in urban and rural population growth have also shifted consumption patterns of wheat, and rising incomes and urban expansion have reduced per capita wheat consumption. However, the impact on diversity may be less straightforward, if urban demand is differentiated across categories of wheat. An increased demand for high-quality wheat products could induce a more diversified wheat production to meet the demands of products for specific niche markets and may well have a positive impact on wheat diversity. The emergence of rural food markets will also affect the demand for diversity. As rural markets improve, they will increasingly shift food consumption patterns in favor of meat, fish, and fruit, goods that have been purchased primarily from the market (Huang and Rozelle 1998). While there is a significant negative relationship between the wheat consumption of farmers and rural market development, demands for more diversified wheat varieties and quality also are expected to increase with rural food market development.

Foreign trade

Foreign trade liberalization is also likely to have an indirect impact on wheat diversity. Wheat is one of the leading importable commodities in China. As discussed in the previous section, China imports mainly high-quality wheat for special uses in the food industry. If trade liberalization leads to an increase in high-quality wheat imports, the imported wheat may substitute for part of the domestic production of this high-quality wheat, and as such reduce the range of wheat varieties in China. However, policy makers could respond to the increasing imports of wheat by increasing wheat quality improvement research, such that high-quality wheat could be produced in China at a more competitive market price.

Overall impacts

There have been myriad changes in agricultural policy and the wheat economy more generally in China in recent decades. Many of those policies and farm-specific factors have simultaneously affected wheat diversity outcomes at the farm field level. Certain factors and policies have increased the concentration of leading cultivars in specific areas; others have held back this concentration. Thus, the overall impacts of agricultural policies and wheat economy changes on wheat diversity are mixed. Impacts of specific agricultural and agriculturally-related policies on wheat diversity require empirical testing with available data to identify the impacts and their causes more clearly.

REFERENCES

De Brauw, A., J. Huang, and S. Rozelle. 2004. The Sequencing of Reform Policies in China's Agricultural Transition. *The Economics of Transition* 12: 427-465.

Fan, S. 1991. Effects of technological change and institutional reform on production growth in Chinese agriculture. *American Journal of Agricultural Economics* 73: 266-275.

Fan, S. and P. Pardey. 1995. *Role of Inputs, Institutions, and Technical Innovations in Stimulating Growth in Chinese Agriculture*. Working Paper, International Food Policy Research Institute, Washington, D.C.

Gale, F., B. Lohmar, and F. Tuan. 2005. *China's New Farm Subsidies*. Electronic Outlook Report from the Economic Research Service. WRS-5-1. Washington, D.C.: U.S. Department of Agriculture.

Hu, R., J. Huang, S. Jin, and S. Rozelle. 2000. Assessing the contribution of research system and CG genetic materials to the Total Factor Productivity of rice in China. *Journal of Rural Development* 23: 33-70.

Hu, R. 1995. China's seed industry and technological change. Unpublished Ph.D. dissertation, Department of Agricultural Economics, Zhejiang Agricultural University, Hangzhou, China.

Huang, J. 2000. *Agricultural Policy and Food Security in China*. Working Paper, Center for Chinese Agricultural Policy, Chinese Academy of Agricultural Sciences, Beijing.

Huang, J. 2001. *Current Reform on Grain Marketing System: Evidences and Implications*. Working Paper WP-00-C1, Center for Chinese Agricultural Policy, Chinese Academy of Sciences.

Huang, J., M. Rosegrant, and S. Rozelle. 1996. *Public Investment, Technological Change, and Reform: Comprehensive Accounting of Chinese Agricultural Growth*. Working Paper, International Food Policy Research Institute, Washington, DC.

Huang, J., S. Rozelle, and M. Chang. 2004. Tracking Distortions in Agriculture: China and Its Accession to the World Trade Organization. *The World Bank Economic Review* 18: 59-84.

Huang, J. and S. Rozelle. 1995. Environmental stress and grain yields in China. *American Journal of Agricultural Economics* 77: 853-864.

Huang, J. and S. Rozelle. 1996. Technological change: Rediscovering the engine of productivity growth in China's agricultural economy. *Journal of Development Economics* 49: 337-69.

Huang, J. and S. Rozelle. 1998. Market development and food demand in rural China. *China Economic Review* 8: 200-220.

IMF (International Monetary Fund). IMF Database. Accessed 2002.

Jin, S. 1997. Determinants of Agricultural Research Funding in China. Master's Thesis, Department of Agricultural Economics, Rutgers University, New Brunswick, NJ. 1997.

Jin, S., J. Huang, R. Hu, and S. Rozelle. 2002. The creation and spread of technology and Total Factor Productivity in China's agriculture. *American Journal of Agricultural Economics* 84: 916-930.

Lin, J.Y. 1992. Rural reforms and agricultural growth in China. *American Economic Review* 82: 34-51.

Liu, X. 1992. Irrigation Investment in China. Unpublished Doctoral Dissertation, Department of Agricultural Economics, University of Philippines, Los Banos.

McMillan, J., J. Walley, and L. Zhu. 1989. The impact of China's economic reforms on agricultural productivity growth. *Journal of Political Economy* 97: 781-807.

MWR (Ministry of Water Resources). 1988-2000 (various issues). Water Conservation Yearbook of China. Beijing: China's Water and Electric Press.

NSBCa (National Statistical Bureau of China). 1980-2002 (various issues). *Zhongguo Tongji Nianjian* [*Statistical Yearbook of China]*. Beijing, China: China Statistical Press.

NSBCb (National Statistical Bureau of China). 1992. Forty Years of Statistics of China's Market Administration and Management. State Statistical Bureau Press: Beijing.

NSBCc (National Statistical Bureau of China). 1995-1999 (various issues). China Ministry of Domestic Trade Yearbook. State Statistical Bureau Press: Beijing, China.

Peng, K. and Z. Xu. 1993. The problems existing in land environment of China: protective countermeasures. *Ecological Economics* 5: 18-22.

Qiao, Fangbin. 1997. Evolution of Forestry Property Rights and Forestry Development. M.S. thesis. Center for Chinese Agricultural Policy and Graduate School, Chinese Academy of Agricultural Sciences, Beijing, China.

Rozelle, S., J. Huang, E. Meng, R. Hu, and S. Jin. 2003. *Diversity of Productivity: The Case of China's Wheat*. Working Paper, Department of Agricultural and Resource Economics, University of California, Davis.

Rozelle, S. and J. Huang. 2000. Transition, development and the supply of wheat in China. *Australian Journal of Agricultural Economics* 44: 543-571.

Rozelle, S., A. Park, J. Huang, and H. Jin. 1997. Liberalization and rural market integration in China. *American Journal of Agricultural Economics* 79: 635-662.

Rozelle, S., E. Meng, S. Jin, R. Hu, and J. Huang. 2000. *Genetic Diversity and Total Factor Productivity: the Case of Wheat in China*. Working Paper WP-00-E9, Center for Chinese Agricultural Policy, Chinese Academy of Sciences.

Rozelle, S., C. Pray, and J. Huang. 1996. Agricultural research policy in China: Testing the limits of commercialization-led reform. *Comparative Economic Studies* 39: 37-71.

Sicular, T. 1995. Redefining state, plan and market: China's reforms in agricultural commerce. *China Quarterly* 143: 1020-1046.

SPB [State Price Bureau]. 1978-2000. *Quanguo nongchanpin chengben shouyi ziliao huibian* [National Agricultural Production Cost and Revenue Information Summary—in Chinese]. China Price Bureau Press: Beijing, China,

Stone, B. 1988. 1993. Basic Agricultural Technology Under Reform. In *Economic Trends in Chinese Agriculture: The Impact of Post-Mao Reforms*. Y.Y. Kueh and Robert F. Ash (eds.) Oxford, England: Clarendon Press.

Wang, J., J. Huang, and S. Rozelle. 2000. An innovation of property right in the groundwater irrigation system and its theoretical explanation. *Economic Research Journal* 4: 66-74.

Wang, L., R. Hu, J. Huang, and S. Rozelle. 2001. Soybean genetic diversity and production in China. *Scientia Agricultura Sinica* 34: 604-609.

World Bank. 1996. *China's Seed Industry: A Sector Report*. China Division, Agriculture Department. World Bank: Washington, D.C.

World Bank. 2001. *The Alleviation of Poverty in China*. World Bank Report. World Bank: Washington, D.C.

Ye, Q. and S. Rozelle. 1994. Fertilizer policy in China's reforming economy. *Canadian Journal of Agricultural Economics* 42: 191-208.

Chapter 4

Policy Influences on Wheat Genetic Diversity in Australia

D. Godden and J.P. Brennan[1]

ABSTRACT

Government policy has had a major impact on the marketing of Australian wheat and on research and development related to wheat. The effects of government policy on the supply of and demand for genetic diversity in Australian wheat production are diverse and often subtle. However, since 1950, there have been three broad periods of policy regimes that affected varietal diversity. The first, up to 1971, was characterised by regulated marketing, wheat breeding in the public sector, and wheat graded as "Fair Average Quality." The second, from 1972-89, was characterised by the continued dominance of regulated marketing and public wheat breeding, but with multiple grades and differing prices. The third period, since 1990, has been characterised by market deregulation, more specialised grades and payments for protein, increasingly commercialised breeding, and increased numbers of varieties released. These broad periods define times at which various government and industry policies, such as research and development and wheat marketing in Australia, have influenced the genetic diversity of the wheat varieties grown in both positive and negative directions since 1950.

Since 1950, government policy has had a major impact on both the marketing of Australian wheat and its research and development process. The policy environment affects the economic and social conditions in which farmers and plant breeders make decisions about the development and use of varieties. These changes directly and indirectly influence the decisions that these groups make about wheat varieties and consequently, the diversity of the set of available varieties. In this chapter, we examine the possible impact of policy changes on decisions about the development and use of wheat varieties and associated levels of diversity.

---

[1]*This chapter draws heavily on Godden and Brennan (2002).*

Genetic diversity in the Australian wheat industry is of significance for three principal reasons. First, given that individual farmers face an array of risks and uncertainties, including price and production risk, the choice of wheat variety offers some opportunities to manage risk and uncertainty in wheat production. Each variety in a set of modern wheat varieties, because of its unique genetic composition, responds differentially to climatic and other environmental conditions. For example, some wheat varieties are optimally sown earlier in the season than others, and all will vary in their level of tolerance to biotic and abiotic stresses. The availability of different varietal types allows farmers to exploit different conditions as they occur across time and space. This factor might be denoted "routine" risk and uncertainty.

Second, plant breeding is an economic activity in the sense that breeders are continually faced with trade-offs within an array of multiple plant breeding objectives. Cultivar improvement within the constraints of available financial resources and within the constraints of what is genetically possible requires complex choices. The priorities determined by breeders—for example, genetically diverse varieties, higher yield, improved quality—govern the array of varieties available to farmers to manage their production systems.

Third, genetic diversity plays a potentially important role in the ecological sustainability of the wheat production industry. The possibility of major breakdowns in the disease resistance of a crop was recognised following the 1970 Southern Corn Leaf Blight in the USA. Without perfect knowledge about the response of each genotype to current and future environmental conditions, there will always be some residual uncertainty about the production stability and sustainability of the existing range of varieties and nearly-available varieties. Crop genetic diversity assists in managing this uncertainty.

Government policy can shift the supply and demand curves for genetic diversity in the Australian wheat industry, but to test the significance of that role would require some consistent representation of "policy." A prerequisite to the construction of such a policy measure is an analysis of the government policies that are most likely to affect the supply and demand for genetic diversity. Godden and Brennan (2002) provide a detailed review of the development of the Australian wheat industry at the colony and state levels to the mid-20th century, as well as the evolution of interventions in the industry and patterns of production for the latter half of the 20th century.

In this chapter, we focus on the evolution of the Australian wheat industry over 1950-2000 and on the concurrent development of the Australian agricultural policy environment. We describe changes in the general policy environment and highlight policies implemented during that period that likely influenced the development and use of genetic diversity in the Australian wheat industry. We then analyse in greater detail possible effects of specific policies on wheat genetic diversity.

Wheat Production in Australia

The development of the Australian wheat industry in the second half of the 20th century is outlined in Godden and Brennan (2002), and further detail on yield and area changes over that period is provided in Brennan and Quade (2000). The area sown to wheat in Australia has more than doubled since 1950, with the expansion taking place mainly in drier environments, especially Western Australia, New South Wales, and Queensland. Western Australia in particular has seen consistently strong growth in wheat area over the past 50 years and has now become the leading wheat-producing state. National wheat yields have increased an average 1.0% per year since 1950 (Brennan and Quade 2000). In the main wheat-producing states, yields have increased most rapidly in New South Wales and Western Australia. Periods of stagnation in yield growth have been observed in some states, but the 1950s and the 1990s were periods of particularly strong yield growth. Overall wheat production in Australia increased by 360% during that period, a compound rate of 3.1% per year (Brennan and Quade 2000).

A key issue in the Australian wheat industry has been the variability in production, due particularly to variable productivity, and attempts to overcome it by industry stabilisation schemes. Godden and Brennan (2002) provide a detailed discussion of the sources of variability in Australian wheat production and some background to the issue of stability and industry stabilisation policies.

The structure of the Australian wheat farms reflects in part the role that variability has played (Godden and Brennan 2002). Wheat in Australia is mainly produced on either mixed cropping farms (particularly in the northern wheat producing area) or mixed cropping-livestock farms (in south-eastern and western Australia). The implication of this mixed farming for genetic diversity is that wheat farmers who wish to decrease the level of risk in wheat production are more likely to do so by incorporating another crop or a livestock enterprise, rather than by growing multiple, genetically-distinct wheat varieties. Australian wheat farms are generally owner-operated family farms with an average wheat area of 1,500 hectares. They are highly mechanised and rely heavily on herbicides. Each farmer generally grows 2-3 wheat varieties, which may be genetically similar, per year. Off-farm earnings that provide a financial supplement and a source of income stability are another means by which Australian wheat farmers manage the effects of variability in production

The Policy Environment

The complex interactions between the wheat industry and the broader Australian economy imply that evaluation of the policy context of wheat production and variability—and thus the role of genetic diversity in managing risk and uncertainty—cannot simply be confined to the wheat industry itself (Godden and Brennan 2002). A specific agricultural industry such as wheat is "nested" within the agricultural sector, which itself is nested within near-agricultural sectors (e.g., other natural resource industries and input-supplying and farm-output-using industries), that are themselves embedded in the national economy. Each of these levels uses inputs supplied by itself and other sectors, produces outputs that are used by itself and other sectors, and connects to the rest of the world via exports and imports. The successive levels imply a complex series of inter-industry and inter-sectoral relationships which may be represented by a general equilibrium model of the economy. Most importantly, each of these relationships is affected by government policy of both economic and non-economic character.

There is a wide range of possible ways in which government policy can affect genetic diversity. Governments can directly influence marketing systems, farm management, and plant breeding objectives, all of which affect spatial diversity (see Chapter 8). Wheat breeding has largely been a publicly-funded activity, farm management has been strongly influenced by government extension activity and by direct financial measures such as taxation, and the wheat marketing system was a statutory system from 1949 until the late 1990s. In addition, macroeconomic conditions and government attempts at macro- and micro-economic management have effects that flow through to the wheat marketing system and directly to farm management and plant breeding.

Influence of constitutional structure

For agriculture in general and the wheat industry in particular, the Australian constitutional structure and its limitations are of major significance. The power to regulate the wheat industry lies broadly with the states. However, the power to regulate exports and interstate trade lies with the Commonwealth government. Marketing schemes, particularly those involving the pooling of income and disbursement at a common payment rate across individuals, most easily operate via a system of levies and bounties; however, these instruments are the exclusive preserve of the Commonwealth. Thus, a national wheat marketing scheme of this form required cooperation between national and state governments, in the form of complementary Commonwealth-State legislation to create a suitable instrument. It took considerable legislative efforts to construct a relatively robust form of national marketing intervention. Conversely, a national marketing scheme could be dismantled by a single disaffected government. Thus, for example, the Commonwealth government decided in the late 1980s to deregulate the domestic marketing of wheat against considerable opposition from the states and wheat-grower organisations.

Continuing deregulation

Deregulation of the financial sector beginning in late 1983 was the catalyst for a sequence of further deregulations affecting large parts of the Australian economy. Once the financial sector was deregulated, some forms of government intervention became increasingly difficult to manage. Other forms of regulation also became increasingly difficult to justify, such as regulation of the domestic marketing of wheat and the storage, handling, and transport of grain. The broader context for the process of deregulation in the Australian wheat industry during the 1980s is discussed in more detail in Godden and Brennan (2002). Progress towards increasing competitiveness was formalised in the National Competition Policy Agreement of 1995. This legislation also affected the way in which former government agencies could operate in the wheat industry.

The "economic rationalist" agenda included a demand for smaller government that had separate philosophical, financial, and economic dimensions (Godden and Brennan 2002). Focus centered on excessive transaction costs and market failure as the principal justifications for government intervention in markets. Not only was a substantial prima facie case of market failure required to justify government intervention, but all interventions were to be carried out efficiently. These expectations led to a questioning of every form of government involvement in the wheat industry and ultimately to a reduction in the level of that involvement.

Trade and international agreements

In the mid-1980s, agricultural trade reform in the Uruguay Round of GATT, which was concluded in 1994, also became an important issue for the Australian government. The principal features of the agricultural agreement were reductions in farm export subsidies, increases in import market access, including conversion of some non-tariff barriers to tariffs, and reductions in producer subsidies. The sanitary (human and other animal) and phytosanitary (plant) agreements sought to limit the application of quarantine-related measures to real health issues.

Other important elements of national economic evolution in the 1980s and 1990s were Australia's participation in key international agreements. These agreements included an increasing focus by governments and individuals on environmental degradation. Australia participated in the UN Conference on Environment and Development in Rio de Janeiro in 1992 that resulted in international treaties on global warming and biodiversity. Both topics had serious implications for Australia. The Trade Related Aspects of Intellectual Property Rights Agreement, one of several attempts to harmonise intellectual property regimes internationally, and the rapid privatisation of the plant gene pool have had significant implications for plant breeding in small countries like Australia.

By the early 1980s, widespread concern was being registered over sustainability issues, and governments generally developed policies to promote and enhance the sustainability of agricultural production (Godden and Brennan 2002). A second aspect of environmental protection was the management of exotic pests and diseases, in particular attempting to control their introduction in an era of increasing travel and greater international trade. The traditional approach to this problem was control via prohibition. However, maintenance of a prohibition system became increasingly costly as travel increased and more entry ports opened. Prohibition as a trade protection measure also came under increased international pressure following the Uruguay Round of GATT. Meanwhile, although concern about the state of natural resources originated in isolated debates about the protection of particular environmental resources, this concern evolved into a wider concern about the general state of environmental resources (Godden and Brennan 2002).

Property rights

Concurrent with the changes in attitudes towards government and social awareness, two major changes in property rights occurred as of the late 1980s. The first involved the recognition of "native title" rights (Godden 1997; Godden 1999). The second was the federal government's enactment in 1987 of a specific law for intellectual property rights (IPR) relating to plant varieties. This law—initially called Plant Variety Rights and, in a major 1994 revision, Plant Breeder's Rights (PBR)—had major ramifications for Australian agriculture, as most varieties grown in broadacre agriculture had been publicly bred. Partly because of temporary inhibitions in applying PBR to all categories of plants and partly because of initial reluctance to apply PBR for varieties that had been produced using growers' funds for agricultural research and development corporations, the effects of PBR on broadacre agriculture were delayed.

Implications for Wheat Genetic Diversity

The implications of many of the macroeconomic and social changes described were that governments at both state and federal levels were slowly reducing the degree of government intervention in the general economy, including the wheat industry. However, the effects of these reductions on the demand for genetic diversity in wheat as a risk management strategy are not unambiguously positive or negative. We now focus on the direct and indirect effects of government policy on genetic diversity in wheat production. An organizing framework for the analysis is provided by de Janvry's (1978) description of a system providing technological and institutional innovations modified by Godden (1997) to incorporate private sector agricultural research. Applying that framework to diversity in recent decades in the wheat industry suggests a complex web of detail, with changing policy decisions having possibly confounding effects. The key policy areas and their impact on the levels of genetic diversity in the Australian wheat industry are summarised in Table 4.1.

Table 4.1 Key policy implications for wheat genetic diversity in Australia.

Policy component	*Effect on diversity (+ increase, – decrease)*
1. Land tenure and property rights	
1a. Native title	?
1b. Land clearing controls	?
1c. Water rights	?
1d. Plant variety/breeder rights	+/–
2. Role of new technologies	
2a. Availability of new genes for disease resistance	+
2b. Semi-dwarf wheats	+/–
2c. Mechanisation allowing more marginal areas to be cropped	+
2d. Chemicals for weed control	+/–
2e. Clover-ley farming systems	+
2f. Fewer, larger farms	+/–
2g. Alternative crops for rotation with wheat	–
3. Marketing structure and prices	
3a. Wheat industry stabilisation schemes	?
3b. Role of pooling for marketing	–
3c. Change in classification system from FAQ to ASW, etc.	+
3d. Price differentials for end-use quality	+
4. Politico-bureaucratic structures	
4a. Dominance of Australian Wheat Board	–
4b. State-based handling and storage systems	–
4c. Coordinating role of GRDC	+
4d. Short-term funding arrangements through GRDC	–

Socio-economic Structure

Land tenure and property rights

The principal change affecting land tenure in the second half of the 20th century was the Australian High Court's native title decisions in the 1990s and consequent legislation. These tenure changes had, however, little impact on wheat production, as native title decisions initially affected only Crown lands and subsequently pastoral leasehold, neither of major significance for wheat production.

Some attenuation of freehold tenure began to develop in response to concerns about the effect of land clearing on biodiversity maintenance and degradation of terrestrial carbon sinks. These changes also had little impact on principal cropping areas, because land was either continuously cropped or in rotation with sufficiently short pasture phases as not to be affected by land clearing controls. Cropping areas likely to be affected by clearing controls were in marginal cropping areas with opportunity cropping, and these changes were unlikely to translate into significant effective changes in the demand for wheat genetic diversity.

Other changes in natural resources property rights included strengthening property rights associated with irrigation water, but again wheat production was unlikely to be affected, as only a small proportion of the crop is produced under irrigation.

Intellectual property rights

There are several possible implications of PBR for genetic diversity. The first set stems from the fact that PBR were intended to encourage private plant breeding. Even if this private stimulus occurred, it may have replaced public breeding in some cases, rather than simply augmenting existing breeding efforts. Alternatively, the stimulus to private breeding may have encouraged public breeders to shift away from finished varieties towards more basic germplasm evaluation activities. Another possible implication is that, even where the private stimulus occurs, additional private breeding tends to encourage largely cosmetic breeding (Godden 1987): that is, similar advances in plant breeding occur as would have without PBR, but competition in the private sector results in morphologically differentiated but genetically similar varieties. Finally, it is possible that PBR increased concentration in the plant breeding industry (Godden 1987). This concentration especially facilitates non-price competition, which may or may not increase numbers of released varieties and/or genetic diversity in these varieties. If a positive effect does exist, it could be the horizontal integration in agricultural input supplies facilitated by intellectual property rights—say, among firms supplying new varieties and agricultural chemicals—and decreased incentives to draw on genetic diversity to control damage from pests and diseases, where chemical solutions to these problems exist.

However, because the implementation of PBR for wheat breeding occurred contemporaneously with the breakdown of the regulated wheat marketing environment, where varietal development and release had been managed as one component of the managed wheat market, the effects of PBR are likely to be strongly confounded with those from wheat market deregulation.

State of technology

In the second half of the 20th century, the state of technology in wheat production evolved in four principal phases (Godden 1999). In the first phase, genetic resistance to major diseases, especially the rusts, became effective across the entire wheat crop. In the second phase, to about 1980, the scale of farm machinery increased dramatically, as tractor size increased and was accompanied by technical innovations such as hydraulics. In the third phase, semi-dwarf wheat varieties largely replaced taller varieties. In the fourth phase, mechanical cultivation for weed control was increasingly replaced by herbicides. The second and fourth of these phases were primarily imported technologies, as Australia had lost its earlier comparative advantage in agricultural machinery innovation and had never developed a significant chemical industry. The chemical revolution, however, was modified by domestic policy considerations regarding occupational health and safety, increasing awareness of negative environmental effects, and eventual concern about developing resistance to herbicides. While the germplasm enabling both the disease resistance and semi-dwarf advances was imported, its transformation into commercial varieties was substantially influenced by the predominantly public domestic plant breeding and research funding institutions.

An additional technological factor, in part stimulated by the wheat industry's crisis at the end of the 1960s, was the search for alternative dryland cropping enterprises. While only a small proportion of possible alternative species ultimately proved widely successful—for example, canola in eastern Australia and lupins in Western Australia—these species proved to be both substitutes to wheat and valuable species in rotations with wheat.

The effects of this technological evolution on the supply of and demand for wheat genetic diversity and on production variability were understandably complex. Elements to be considered include:

- Both the rust resistance and semi-dwarf phases increased genetic diversity, in the narrow sense that additional, specific genes were incorporated into commercial varieties to express these particular characteristics. The extent to which Australian wheat breeders considered genetic diversity in their activities and the mechanisms they utilised is reported in Chapter 5. Brennan et al. (1999) describe a first attempt to model Australian farmers' demand for genetic diversity in wheat production.
- Large-scale machinery improved the timeliness of operations, enabling production in more marginal areas and increasing the demand for a greater range of cultivars with new qualities (Whitwell and Sydenham 1991). The expansion in cultivated area increased opportunities for natural selection of diseases and thus implicitly increased the demand for genetic improvement, if not greater diversity. However, the relative homogeneity of the new wheat lands and the large scale of operations may have led to a low demand for genetic diversity *within* these new areas.
- Improved chemicals similarly enabled expansion into more marginal areas and enabled larger cropping areas, with similar effects as those from machinery. New chemicals (e.g., for weed control) also allowed previously high-cost crops such canola to become more financially competitive with existing enterprises.
- The clover-ley farming revolution of the mid-20th century in southern Australia ultimately encouraged development of acid soil lands and, as a consequence, the demand for wheat varieties tolerant of less-favourable soil conditions.

Partly induced by the above changes, the average size of wheat farms grew by 2% per annum during 1967-87 (Whitwell and Sydenham 1991). As regards the demand for crop diversity, the implications of these changes in farm size and the concentration of production in a small number of farms are ambiguous. Smaller-scale farmers are more likely to be risk averse, but it is not clear in the case of wheat production whether or not this risk aversion translates into a demand for greater crop diversity. Small farms are relatively more dependent on off-farm income (Godden and Brennan 2002) and are more likely to have relatively higher costs of accessing and managing greater genetic diversity. Thus, small-scale farmers may have a lower effective demand for crop diversity.

Marketing structure

From a marketing policy perspective, the wheat industry worked under a series of generally five-year plans, from 1948-49 to 1987-88. Whitwell and Sydenham (1991, p. 134) summarised previous analysis of the objectives of the wheat marketing legislation as follows:

- *With respect to income* — "to increase and secure the standard of living of wheat farmers, to maintain comparability between farm and non-farm incomes, to assist low-income producers, and to stabilise farm incomes";
- *With respect to price* — "to guard against 'ruinous' prices, to generate prices fair to producers and consumers, to avoid excessive fluctuations in prices, and to provide 'orderly marketing' (that is, to moderate the forces of economic competition between producers)";
- *With respect to production* — "to produce enough wheat to meet domestic requirements, to stimulate export production, to encourage efficient production, and to orient production towards more-favoured areas";
- *With respect to national policy* — "to earn more export income, to constrain the federal government's fiscal liability, and to encourage the development of rural areas".

The means by which these objectives were initially pursued were a guaranteed minimum price, whose starting point was the assessed cost of production, for specified export quality wheat. While there was some re-ordering of objectives over time, the basic structure was resilient (Whitwell and Sydenham 1991) until the early 1980s, when tentative domestic market deregulation commenced, followed by effective domestic market deregulation in 1989. The Australian Wheat Board was privatised in July 1989 to form AWB Limited. The export monopoly was retained for a period of time, but its existence is regularly reviewed under national competition policy guidelines.

There is debate over whether the schemes actually did stabilise prices (Whitwell and Sydenham 1991). However, in the context of within-crop genetic diversity, the average level of the wheat price and its relationship with alternative enterprises are of less relevance than variations in and relativities of the prices of different kinds of wheat. Prior to about 1970, Australian wheat was sold largely under the "Fair Average Quality" (FAQ) system. In that system, all wheat grown in a given area was mixed and a weighted average sample was declared FAQ for that season and region, and all wheat received the same FAQ price. While it did not explicitly encourage within-field diversity, this system did not penalise within-field diversity, as long as the heterogeneity was within the limits of the local FAQ declaration (Whitwell and Sydenham 1991).

However, even with the FAQ system, buyers were aware of varying characteristics of wheats sourced from different areas and purchased accordingly. That is, if buyers knew where to source "premium" quality wheats, they could do so at FAQ prices. Nevertheless, in the prime hard wheat growing areas of northern NSW and Queensland initially, and subsequently elsewhere, there were opportunities to benefit from partial segregation of premium quality wheats, and thus encouragement of varietal specialisation. To the extent that this varietal specialisation occurred, it represented a greater diversity of *cultivars* of wheat grown, and the move to greater segregation encouraged greater genetic diversity (Godden and Brennan 2002).

However, a force working in the opposite direction was the increasing demand for uniformity in wheat batches (Whitwell and Sydenham 1991). This demand for uniformity imposed greater pressures on the grain handling system to increase segregation of different quality wheats. Upon confirmation that segregation was in fact possible in the handling and storage system, greater demands for uniformity in farmer deliveries followed. This latter demand led to an increasing demand for within-crop uniformity and a reduced tolerance of within-crop diversity. Thus the move from FAQ to increasingly tight specifications of class or grade is likely to have decreased within-field diversity while encouraging the diversity of varietal types between regions, as they increasingly specialised in the production of types most suited to the local environment.

Relative prices

Changes in the price of wheat relative to the output prices of other enterprises that could be undertaken on wheat farms led to substantial, and often rapid, switches in farm output. These switches indicate that a "wheat" farmer's principal form of defence against price variability was through enterprise diversification, especially until the 1970s. There was a substantial switch into wool from wheat when wool prices rose in the 1950s, and back again into wheat in the 1960s. The increased specialisation in wheat production constrained the availability of livestock enterprises, as biological or economic complements to wheat production. Other cropping enterprises, especially the development of oilseed production stimulated by the over-production of wheat in the late 1960s (Whitwell and Sydenham 1991), remained as a form of defence against price variability. The lower the correlation between the wheat price and prices of these other grains/oilseeds, the more effective the stabilisation. Other cropping enterprises did not, however, provide much defence against rainfall-induced variability in wheat yields or production, with the possible exception of summer-growing crops.

Politico-bureaucratic Structure

Social pressure system

The FAQ scheme reflected a more deep-seated attitude to the wheat industry than simply a wheat pricing mechanism. FAQ reflected a social pressure system that emphasised egalitarianism, also represented by attempts to even out returns over space through grain pooling and cost averaging, and across time through the stabilisation fund. That egalitarianism diminished over time, both as egalitarianism diminished in the wider Australian community and within the wheat industry itself. The egalitarianism in general represented an external attempt to manage a risky environment. To the extent that it was successful, farmers would have been less reliant on internal risk management mechanisms. For example, state-wide cost averaging in wheat pools until 1978 favoured growers more distant from domestic markets or seaboard terminals. To the extent that more distant growers were on the drier margins, cost averaging encouraged increased production variability and possibly the demand for varieties more suited to the drier margins. Demand for greater overall genetic diversity may have thereby increased through demand for cultivars suitable for these growing environments.

Research levies

Beginning in the mid-1950s, the wheat industry in concert with government established wheat industry research funding arrangements based on a production levy and matching grants from Commonwealth consolidated revenue (Brennan and Mullen 2002). This funding supplemented the core funding of research activities by public sector institutes to support a wide range of wheat research, including plant breeding. In 1989, the Grains Research and Development Corporation (GRDC) replaced the existing arrangements for wheat research and also incorporated the previously separate crop funding arrangements for other grains. The GRDC has been very proactive in managing its research portfolio and has also been active in consolidating its research portfolio for wheat breeding in particular. The impact of the public funding of plant breeding since the mid-1950s for genetic diversity is unclear. However, as indicated in Chapter 5, Australian wheat breeders reported that changes to the wheat research funding arrangements in the 1990s have favoured short-term outcomes over longer-term breeding strategies focused on or leading to genetic diversity.

Innovation Production

Public sector

Up to the 1970s, a generally buoyant attitude to agricultural research prevailed. With expectations for positive marginal returns from additional research expenditures, wheat breeding programs were well maintained. While there were critics of the efficacy of wheat breeding (Campbell 1977), the rate of progress in yield for varieties in comparable classes of wheat were similar to the UK (Godden and Brennan 1994), where national average yields were increasing much more rapidly. Institutional constraints on the types of varieties that would be accepted by the Australian Wheat Board limited breeders to developing and releasing wheats for human consumption. Research on feed wheats, where many of the large advances in yield gains were taking place, was thus restricted.

The general constraints on government expenditure noted above began to constrain research activities from the late 1970s. Partly as a consequence of funding constraints, and partly exogenously, government research organisations increasingly developed formal, integrated research planning and management mechanisms. These management and funding changes reinforced changes occurring in the research funding bodies during the 1980s, which had been stimulated by the new research funding arrangements under the Primary Industries and Energy Research and Development Act of 1989. The aggregate effect of these changes was likely to have reinforced the general impression reported by Australian plant breeders of increasing constraints on the type of work that might lead to increased genetic diversity (see Chapter 5).

Private sector

Prior to the Plant Variety Rights Act of 1987, private plant breeding in Australia had a limited history of one firm primarily focused on the development of F1 hybrid varieties. However, at that time there still had been no successful releases of commercial hybrid varieties. In contrast, between the first application for PBR for wheat in 1991 and 1999, PBR was granted in Australia for 37 wheat varieties. Public breeders have increasingly used companies and/or joint ventures to market their varieties, possibly to increase the effectiveness of marketing, possibly to quarantine revenue from normal funding processes, and possibly to protect the public organisation from litigation in case of disputes. More recently, intellectual property rights for new wheat varieties are held jointly with GRDC. Australian wheat breeders report that they consider that genetic diversity has not been affected by PBR (Chapter 5).

Conclusions

Since 1945, government policies have had a major impact on the marketing of Australian wheat and on research and development. It is probable that the dominant impact would come from industry policy specific to wheat, but the importance of agricultural exports in the early part of the post-war period and the importance of wheat exports within agricultural exports as a whole suggest that — for part of the period at least — wheat industry policy is likely to have been a part of macroeconomic policy. The dominance of government policy over the wheat industry in the first half of the post-war period thus creates some difficulty in discerning the influence of specific wheat industry policies on genetic diversity. While the importance of agriculture in the macro-economy has declined over the post-war period, the emphasis on reducing government intervention in the economy in the second half of the period nevertheless indicates that the deregulation occurring in wheat policy was part of a much larger policy agenda.

The possible effects of government policy change on the supply and demand for genetic diversity in Australian wheat production are diverse and often subtle. Implemented policies may be synergistic with existing policy or may neutralise, fully or in part, existing policy. It is unlikely that a single "policy" variable could be constructed to represent all the possible effects of government policy on supply of and demand for genetic diversity. However, in the post-war period, there have been three broad periods of policy regimes that affected varietal diversity:

(a) Pre-1971: Characterised by regulated marketing through the Australian Wheat Board; wheat breeding in the public sector; wheat graded as FAQ;
(b) 1972-89: Characterised by the continued dominance of regulated marketed marketing through the Australian Wheat Board; wheat breeding in the public sector; multiple grades with differing prices;
(c) Post-1990: Characterised by market deregulation and the privatised AWB Limited; more specialised grades and payments for protein within grades; increasingly commercialised breeding influenced by the role of the GRDC; and increased numbers of varieties released.

These broad periods define times at which various government and industry policies have had varying influences on genetic diversity in Australian wheat. It is likely that, with the current rapid rate of policy change, further influences will take place in the future.

REFERENCES

Brennan, J.P., D. Godden, M. Smale, and E. Meng. 1999. Variety choice by Australian wheat growers and implications for genetic diversity. 43rd Annual Conference, Australian Agricultural and Resource Economics Society, Christchurch, New Zealand.

Brennan, J.P. and J.D. Mullen. 2002. Joint funding of agricultural research by producers and government in Australia. In *Agricultural Research Policy in an Era of Privatization*, D. Byerlee and R.G. Echeverria (eds). CABI International, Wallingford, 51-65.

Brennan, J.P. and K. Quade. 2000. Longer-term changes in Australian wheat yields. *Agricultural Science* 13: 37-41.

Campbell, K.O. 1977. Plant breeding to what purpose. *Journal of the Australian Institute of Agricultural Science* March/June: 48-52.

de Janvry, A. 1978. Social structure and biased technical change in Argentine agriculture. In *Induced Innovation: Technology, Institutions, and Development,* H.P. Binswanger and V.W. Ruttan (eds). Johns Hopkins University Press, Baltimore.

Godden, D. 1987. Plant variety rights: Framework for evaluating recent research and continuing issues. *Journal of Rural Studies* 3: 255-272.

Godden, D. 1997. *Agricultural and Resource Policy: Principles and Practice,* Oxford University Press, Melbourne.

Godden, D. 1999. A Century of Agricultural Progress in Australia. 43rd Annual Conference, Australian Agricultural and Resource Economics Society, Christchurch, New Zealand.

Godden, D. and J.P. Brennan. 1994. Technological change embodied in southern N.S.W. and British wheat varieties. *Review of Marketing and Agricultural Economics* 62: 247-60.

Godden, D.P. and J.P. Brennan. 2002. Policy influences on genetic diversity in Australian wheat production. *Australian Agribusiness Perspectives,* No. 52, <www.agrifood/Review/Perspectives/Godden>

Whitwell, G. and D. Sydenham. 1991. *A Shared Harvest: The Australian Wheat Industry, 1939-1989,* Macmillan Australia, Melbourne.

Chapter 5

Breeder Demand for and Utilization of Wheat Genetic Resources in Australia

J.P. Brennan, D. Godden, M. Smale, and E. Meng[1]

ABSTRACT

Australian wheat breeders were surveyed to assess the importance they gave to genetic diversity and identify their key issues of concern relating to crop diversity. Areas addressed included breeders' attitude toward diversity and the diversity available in their current "gene pool" (that is, the breeding materials they use and/or have access to). The sources of materials that breeders use to maintain and/or enhance diversity in their programs are identified, and the ways in which diversity influences breeding decisions are examined. More importantly from the policy viewpoint, the survey responses identify changes in the environment in which breeders operate that affect their ability to enhance crop diversity. Funding constraints, in particular, are shown to influence the extent to which breeders can utilise genetic diversity. The findings raise important issues concerning the future genetic diversity of Australia's wheat, as well as the extent to which economic problems related to genetic diversity may arise in the future.

The complex issues related to the supply and demand for crop genetic diversity at the household and aggregate levels have been increasingly addressed in the literature (Smale 2006; Smale et al. 2003; Bellon 2004; Smale 1998). The spatial and temporal distributions of varieties across Australian wheat-producing areas represent the simultaneous outcome of the supply of genetic diversity from the nation's wheat breeding programs and the demand for diversity, as expressed through the demand for varieties by the industry. It is crucial to know the range and level of diversity incorporated in the varieties developed and released by breeding programs in order to analyse diversity

[1] *This chapter draws heavily on Brennan et al. (1999).*

outcomes. More specifically, the role of specific breeding objectives within breeding programs in determining germplasm use has been linked to diversity outcomes over time. Diversity was greater in CIMMYT wheat cultivars developed during periods in which increasing diversity was included as an explicit breeding objective (Warburton et al. 2006; Reif et al. 2005), reinforcing the importance of breeding programs and the factors that determine breeding priorities.

At a practical level, breeders determine which genetic resources are used to develop varieties. Their role is a key element in explaining past and present trends in on-farm diversity. A clearer understanding of breeders' methods for setting priorities, their perceived constraints, and their perception of their operating environment is therefore important. Farmers demand varieties for yield or other attributes such as quality or disease resistance, but the outcome of their choices in the aggregate, as constrained by the range of varieties available to them, determines the pattern of spatial and temporal diversity.

Information has been obtained on the role of genetic diversity in wheat breeding programs in Australia through a survey of wheat breeders (Brennan et al. 1999). In particular, information was sought on the importance breeders placed on genetic diversity and the extent to which policy-related factors can inhibit or assist the broadening of that diversity. That survey covered breeders' perceptions of the supply of genetic diversity and the factors that shift or influence that supply. Other surveys of breeder utilisation of genetic resources in crop breeding programs include the studies by Duvick (1984) and Rejesus et al. (1996).

Previous work on Australian wheat (Brennan and Fox 1998) found cause for concern that the genetic diversity was narrowing on farms in some eastern Australian states. In general, however, based on an analysis of pedigree-based measures of average and weighted coefficients of parentage (COPs) among modern varieties (Souza et al. 1994), the diversity among the pool of wheat varieties grown in Australia remained high from 1973 to 1993. Consequently, as has been argued by Souza et al. (1994) and Smale and McBride (1996), declines in diversity that result when farmers choose to grow large areas using only a few varieties reflect the factors that influence farmer demand relatively more than the decisions of wheat breeders involved in the development of the supply.

For field crop production in the United States, Duvick (1984) argued that although the crop area was concentrated in a relatively small number of favored varieties, the genetic base of elite germplasm is wider and provides more useful diversity than is often assumed. In addition, relatively frequent varietal replacement among modern varieties creates temporal diversity that substitutes for spatial diversity found where farmers cultivate more heterogeneous landrace populations. Genetic reserves held by crop breeders are also significant. Data from an international survey of wheat breeders conducted by CIMMYT in 1995 (summarized in Rejesus et al. 1996) showed that the crossing blocks in developing countries contained larger sections of landrace materials and lines from CIMMYT international nurseries than those of high-income countries. As a result, their parent material may be more genetically diverse in types and geographical origin.

In the following section, the survey of Australian wheat breeders is described and the results presented. Implications are drawn in the final section.

SURVEY OF AUSTRALIAN BREEDING PROGRAMS

The Survey

In 1998, a survey questionnaire was sent to each of the 14 public and private Australian wheat breeding programs (Brennan et al. 1999). Twelve completed responses were received. A total of 18 questions was asked, grouped as follows: (a) breeder perceptions and attitude toward diversity; (b) assessment of diversity in the current gene pool; (c) sources of materials used to maintain and/or increase diversity; (d) diversity as a priority in breeding objectives; (e) perceived changes over time in diversity levels; and (f) impact of funding constraints on diversity. The results of the survey are described in the following sections.

Attitudes toward Diversity

Breeders were asked if they believed that a lack of genetic diversity available to their program was constraining their progress in breeding. Rather than the *availability* of diversity, which was viewed as being generally adequate, breeders felt the constraints on their breeding progress were due to the lack of *means to use* effectively the genetic diversity available. Thus, the problem lay not in the availability of adequate diversity when specifically sought out, but rather in their inability to exploit that diversity fully because of limited funding.

When they were asked to rate the extent to which progress toward specific goals was constrained by a lack of available diversity (Table 5.1), the breeders gave generally consistent replies.

- Very few breeders considered it a primary constraint to yield progress.
- For rust resistance, breeders were unanimous in their opinion that a lack of genetic diversity was not limiting progress. However, breeders expressed some concern that while they have ready access to the main known rust resistances, sources of effective resistance available worldwide were limited.
- For other biotic and abiotic stresses, breeders ranked the extent to which lack of diversity was constraining progress as medium or high.
- Diversity in quality was generally ranked low or zero as a constraint.

Table 5.1 Breeder ratings on obstacles presented by lack of available diversity*.
(H: high; M: medium; L: low; 0: zero)

	In your collection				*Within Australia*				*Worldwide*			
	H	*M*	*L*	*0*	*H*	*M*	*L*	*0*	*H*	*M*	*L*	*0*
Yield	1	4	5	2	2	4	4	2	2	4	2	4
Rust resistance	0	0	9	3	0	2	8	2	0	4	4	4
Other biotic resistance	2	5	3	2	2	6	2	2	1	6	1	4
Abiotic stress	1	5	3	3	1	6	2	3	1	6	0	5
Quality	0	2	6	4	0	3	5	4	0	2	4	6

* Out of total response of 12 breeders

There was no evidence of regional differences in these responses, at least partly because the small number of respondents made such trends difficult to identify.

Diversity in Current Gene Pool

Diversity in current program

Breeders were asked to describe the diversity of their current gene pool, and generally rated it as "medium" or "medium-high." Overwhelmingly, breeders of commercial varieties consider that they maintain a moderate level of diversity in their own programs.

Entries in crossing blocks

As an indication of the size of the breeding programs, breeders were asked to estimate the total number of entries in their crossing blocks; that is, the potential parental lines in the most recent season. The size of the breeding program does not necessarily carry implications for diversity, since the type of germplasm included in the crossing block as well as the frequency by which new and diverse entries are added over time are also important factors. The number of entries ranged from 40 to 550, reflecting differences in breeding approaches and resources, with a mean of 231 lines for all breeders. About 30% of breeders had 50 lines or fewer in crossing blocks, with a similar proportion having 400 or more.

Types of materials in crossing blocks

Breeders were then asked to specify by source the percentages of entries in crossing blocks, on average over the past five years (Table 5.2). Since the main sources of materials used for crossing are the breeders' own programs, they are for the most part working with adapted materials. When looking outside their programs, Australian breeders generally rely very little on landraces, preferring to use more adapted materials either from other Australian

Table 5.2 Types of materials in crossing blocks (%).

	*Weighted mean**
Landraces	3
Lines from own program	35
Lines from other Australian programs	18
Lines from overseas national programs	10
Lines from CIMMYT or ICARDA	19
Other (mainly specialized germplasm)	15

* Weighted by number of entries in crossing blocks

programs or from international centers such as CIMMYT in Mexico and the International Center for Agricultural Research in the Dry Areas (ICARDA) in Syria. Materials from national breeding programs outside of Australia were used less frequently. Specialized programs such as the durum program used fewer materials from other Australian programs, because they are less relevant to breeding goals for products such as durum breeding. Breeders also turn to materials such as *Triticum tauschii* derivatives, synthetic hexaploids, and specialised germplasm, for specific attributes from a variety of sources.

Materials to Maintain/Enhance Diversity

Introduction of new materials to program

To assess the rate at which breeders routinely seek to introduce sources of genetic diversity from outside their program, the survey asked what percentage of the materials in crossing blocks was new each year. On average, breeders introduced 25% new materials each year. The percentage rises to 29% when weighted by number of lines in the crossing blocks. The percentage was higher for breeders emphasising a germplasm development objective or in a specialized program, such as durum breeding, in contrast to those focusing on the commercial release of mainstream varieties. From these responses, breeders of commercial varieties apparently prefer to work with a known population and to introduce new materials only as sources of particular traits.

Types and sources of materials introduced to increase genetic diversity

Breeders had looked to a range of sources for novel materials and have introduced various types of materials over the previous five years, with the specific objective of increasing the genetic diversity in their programs (Table 5.3). The two main sources of materials have been lines from CIMMYT or ICARDA international nurseries (40% of all materials introduced by the breeders) and varieties or advanced lines from other Australian breeding programs (29%). Other significant types of materials have been varieties or advanced lines from overseas national programs (14%) and materials with specific traits from various public and private sources (18%). The very low

Table 5.3 Types of materials used to increase genetic diversity (%).

	Mean
Landraces	2
Lines from other Australian programs	29
Lines from overseas national programs	14
Lines from CIMMYT or ICARDA	40
Other (mainly specialized germplasm)	15

level of use of landraces (2%) again demonstrates that breeders in Australia prefer to use more adapted materials for introducing diversity. These results are again similar to those reported in Rejesus et al. (1996).

Breeders utilize a wide range of sources to broaden the genetic base of their programs. The main sources are listed here in order of the number of times mentioned:

1. International nurseries, especially from CIMMYT and/or ICARDA;
2. Exotic (non-commercial) lines from germplasm development programs, such as those conducted by various agricultural organizations and universities;
3. Australian Winter Cereals Collection, Tamworth;
4. Lines from other national programs, especially the USA, Canada, and Europe;
5. Lines from other Australian breeding programs;
6. Other lines from CIMMYT, such as synthetic wheats;
7. Overseas private company sources.

Use of material from Australian Winter Cereals Collection

All breeders reported having used the Australian Winter Cereals Collection (AWCC) in the past five years. The average number of requests reported per breeder in that period was 24, or approximately 5 per year. The AWCC is commonly used to obtain specific germplasm with targeted attributes to broaden the genetic base of breeding programs. Despite the relatively low number of requests, the breeders generally judged the material they obtained as "useful," "very useful," or "essential" in enabling them to expand the range of genetic materials required. Analysis based on a survey of requestors of germplasm samples from the U.S. National Plant Germplasm System show that germplasm described as having "useful" accompanying data had higher rates of use in breeding and other research programs. These results hold both for data for the trait of interest and for other data provided (Day-Rubenstein and Smale 2004). Our survey did not obtain information on the type of accompanying, or descriptor, information for the germplasm requested from AWCC, to allow for further exploration of their role in the significance of the AWCC.

Role of Diversity in Breeding Decisions

Effect on crosses made

All breeders indicated that a concern for adèquate genetic diversity influences their decisions on the type of materials used in their crosses. Most crosses are based on adapted germplasm, which leads more directly to varieties for release. However, most breeders indicated that they make some—albeit often a relatively small proportion—of their crosses with more diverse material in wider crosses with the specific objective of adding variability for grain yield, quality, and stress tolerances to their programs.

Effect on crossing schemes used

The need to introduce genetic diversity affects, at least in part, the crossing scheme used. Almost all breeders specifically mentioned that they were more likely to use backcrossing when working with exotic or poorly adapted material to increase the gene frequency of the adapted parent and improve the probability of commercial acceptability. Some breeders also mentioned other crossing methods used with these materials, including bi-parental and triple crosses, as well as specialized crossing strategies utilized for backcrossing more exotic materials.

Effect on selections made

Most breeders indicated that the selections they make in their programs are affected by their concern for genetic diversity. Where backcrossing is used, breeders can make fewer selections, while for other crosses where more distantly related materials ares used, a larger number of selections may be necessary to get useable types. In general, breeders confirmed that heritability and ease of selection, the latter of which is closely related to parental material used, affects selection methods and approaches, with more cycles of crossing/ selection needed for the more diverse material.

Effect on varieties released

A slim majority of breeders indicated that genetic diversity influences their decisions on which varieties to release, as they require a range of maturities, stress tolerances, and quality types in the pool of available varieties. However, several breeders countered that diversity plays no role in those decisions, or that it would only do so if there were a yield advantage. It appears that breeders generally strive to release varieties with a range of genetic backgrounds, but if they find that an advanced line has some yield or quality improvement they will release it, despite any genetic similarity to other existing varieties.

Determinants of genetic materials used in program

When making decisions on the genetic materials to use in their programs, breeders consider the limitations of their current germplasm, the need to maintain variability for important characteristics in their program, the availability of suitable material, and an assessment of the needs of the industry in terms of quality, yield, resistances, and market types. Funding constraints and the immediate expectations of funding bodies and growers were also listed as factors influencing the choice of genetic materials, as a result of pressures from those sources to make fewer wide crosses and focus on specific, shorter-term goals.

Changing Institutional, Technological and Environmental Factors

Breeders were asked whether there had been changes over the past 15-20 years in the environment in which they operated, influencing the extent to which diversity was incorporated into their programs. The majority of respondents described a range of changes in either the extent to which diversity is available or the way in which they incorporated diversity over that period. Factors influencing the extent to which diversity was incorporated in the breeding programs included: (a) changes in market quality parameters; (b) changes in funding levels and arrangements; (c) changes in disease spectrum; (d) changes in the quarantine system; (e) changes in availability of exotic breeding materials; and (f) technological changes.

Impact of changes in market quality parameters

Without a more rigorous analysis on the subject, the overall impact of market quality parameters on genetic diversity remains ambiguous. On the one hand, quality objectives have become more prominent in breeding programs over the past 15-20 years, as markets have become more discerning in the wheat qualities they demand. Because many lines in Australia are now rejected for reasons of unsuitable market quality, some breeders assert that they are more constrained in the choice of parents and consciously limit their use of lines that exhibit poor attributes for important quality traits. On the other hand, the increase in the range of quality types now acceptable to the market permits the release of wheat varieties in several different quality categories. This change has allowed breeders to make broader crosses than if there were only one or two narrowly defined quality types. The result is a set of varieties that are widely diverse in maturity types, dough qualities, and end uses, although not necessarily more genetically diverse.

Impact of changes in funding levels and arrangements

The nature of funding for Australian breeding programs has changed since around 1990 (Brennan and Mullen 2002) from one of core funding primarily from State governments to levy-based funding through the Grains Research and Development Corporation (GRDC). Breeders reported that the funding structure has consequently become more skewed toward short-term outcomes for both germplasm development and variety release, and that the funding structure discourages wide crosses and crosses for parent building due to the unpredictable outcomes of such crosses. This disincentive has led some breeders to emphasize backcrossing for the addition of specific characteristics over crosses aimed at expanding genetic diversity. Exceptions exist: for the durum breeder, for example, funding has become greater and more flexible, allowing the pursuit of small projects outside the main breeding program.

Impact of changes in disease spectrum

Wheat diseases have had an ambiguous impact on genetic diversity over the past 15-20 years. In that time, some wheat diseases have spread through breakdowns in resistance and changes in agronomic practices such as cropping rotations and stubble retention and reduced tillage. Changes in farm management have required the incorporation of resistance to additional diseases into new varieties. At the same time, the introduction of new diseases, such as stripe rust in the late 1970s and a subsequent widespread outbreak in 2004, has reduced diversity due to the need to incorporate genes with resistance to stripe rust into new varieties. The use of the few available resistant genes initially limited the materials that could be used in some programs until effective and broad-based resistance was in place. On the other hand, the achievement of a relatively stable situation over the same period for stem and leaf rust (and until more recently for stripe rust) permitted additional breeding objectives to be addressed. A larger amount of resources could thus be allocated to incorporate additional diversity for other characteristics. Moreover, the increased threats of exotic pests and diseases such as Karnal bunt and Russian Wheat Aphid have influenced breeders to identify resistant materials in sources that were not previously utilized in the breeding programs.

Impact of quarantine system

Breeders also identified the rate of movement of material through quarantine as a constraint to the incorporation of diversity into new materials. During the 1990s, the Australian Quarantine Inspection Service introduced "cost recovery" for the quarantine aspects of importing germplasm into Australia. Breeders introducing germplasm into Australia were therefore faced with higher direct costs than in the past. Given budget constraints, these costs have become increasingly significant, and some breeders reported reductions in the number of lines they import for testing. These decisions could have important impacts on the genetic diversity of Australian crops in the future.

Impact of changes in availability of exotic breeding materials

The recent development of new synthetic hexaploids based on *Triticum tauschii* and their availability through CIMMYT and ICARDA has provided Australian breeders with access to a wide range of new diverse materials with potentially valuable traits such as disease resistances. Recent research confirms the positive effect on wheat diversity of the synthetic hexaploids (Warburton et al. 2006), and they are perceived by many breeders as an interesting new source of available diversity. Germplasm enhancement programs recently established by some organizations in Australia were also mentioned. These programs have become sources of diversity, particularly for certain targeted characteristics such as coleoptile length and rust resistance. In addition, the increased interest in durum wheats and other tetraploids in recent years has increased the amount of genetic information available about durum wheats. These developments have greatly increased the diversity in durum breeding programs in particular.

Impact of technological changes

Breeders commented on several aspects of technological change in the past 15-20 years that have had an impact on genetic diversity. For example, improved communication methods such as internet access have sped up the access to and rate of exchange of information on available diversity. In addition, methods used to incorporate diversity have changed with the development of molecular markers and doubled-haploid techniques, which facilitate the incorporation of certain characteristics into breeding materials. However, the other side of the coin is again the risk that use of doubled-haploids and marker-assisted selection may lead to fewer wide crosses and an increased emphasis on targeted, short-term outcomes. Concern was expressed that the expanding use of these techniques could lead to a future reduction in diversity.

Impact of Funding Constraints on Diversity

Extent of funding constraints

Breeders gave two different responses regarding the impact of recent funding constraints on diversity. A third of the breeders responded that funding was not constraining their programs at present, while one program reported that its funding had in fact increased substantially in recent years. However, most breeders indicated that funding available to them constrained the extent to which they could introduce and use diversity in their programs. Given reductions in real levels of total funding for breeding, breeders have had to establish priorities for the types of diversity they seek out and for the amount of new diversity to introduce and incorporate into their programs each year.

Uses of additional funding

Breeders were asked to prioritize the areas to which they would allocate any additional funding obtained for their program. The responses were extremely variable (Brennan et al. 1999) and are summarized in Table 5.4. Overall, "Improved selection methods, including DNA markers, etc." was ranked as the highest priority. This was closely followed by "Genetic diversity for other biotic resistance" and "Genetic diversity for yield." Most other options relating to genetic diversity were ranked as low priorities, especially those for quality and rust resistance. The reason for the consistently low ranking of "Genetic diversity for quality" was the belief that the existing genetic variation in quality is adequate and that the tight specifications for quality demanded by the market limit the broad incorporation of characteristics currently not present in marketed varieties. Similarly, there was a belief that sources of rust resistance available to the breeding program are also sufficient.

Table 5.4 Ranking of areas for additional funding.
(1 Highest, 10 Lowest)

	Overall ranking
Genetic diversity for yield	3
Genetic diversity for rust resistance	7
Genetic diversity for other biotic resistance	2
Genetic diversity for abiotic stress	5
Genetic diversity for quality	8
Novel sources of diversity	6
Improved selection methods	1
Increased emphasis on quality	4
Increased attention to niche markets	9
Others	10

Taken in conjunction with the expressed concern that the use of genetic markers and marker-assisted selection may result in a reduction in overall diversity, the fact that the highest priority for additional funds was improved selection methods, including DNA markers and other molecular breeding tools, could highlight possible negative impacts on diversity levels in the future.

Implications of Findings

Overall in the survey, Australian wheat breeders expressed cautious optimism in their approach to genetic diversity. Breeders were generally satisfied that they had a reasonable, though not high, amount of diversity in their programs. Few considered that progress with their major breeding objectives was highly constrained by the lack of available genetic diversity. Cautious optimism also characterises Duvick's conclusions in 1984 for the situation in the United States, although the breeders surveyed internationally in 1995 by Rejesus et al. (1996) appeared more sceptical. The concern expressed by breeders in that international survey were likely influenced in large part by the changes in the global political climate and reflected the fear that variety protection laws would reduce access to materials. Australian wheat breeders are less reliant on other countries for the diversity in their materials, largely because of the materials already present in the Australian Winter Cereals Collection at Tamworth.

All breeders introduce new genetic materials into their programs regularly with an average of approximately one-quarter of their crossing blocks each year. That percentage provides some evidence of a continuing new search for useful materials in their programs. While many of those new lines were obtained from other Australian breeding programs, the principal sources of novel materials have been CIMMYT and/or ICARDA nurseries. The importance of the international agricultural research system to the Australian wheat industry, identified in Brennan and Quade (2004), is clearly continuing strongly. The consistent use of lines from CIMMYT, including synthetic wheats, and lines from other national breeding programs in addition to those from other international nurseries underscores the importance to Australia of international linkages in wheat breeding.

A key contribution to the availability and distribution of germplasm for Australian wheat breeders is the Australian Winter Cereals Collection at Tamworth. All breeders reported using the AWCC as a source of materials, and the assessment of the materials they obtained was very positive. It clearly plays a key role in enabling the expansion of the genetic base in Australian breeding programs. In addition to its role as a repository and gene bank for the Australian industry, the AWCC plays a key role in the introduction of materials from overseas and the implementation of some quarantine requirements for that material. There appear to be considerable economies of scale associated with a central agency such as AWCC, including taking a leading role in the introduction of international nurseries.

Breeders reported that the genetic diversity of the materials they worked with influenced activities throughout the breeding cycle in their programs. The crosses made, the crossing schemes used, the selections made, and the decisions on varieties released are all influenced, to a greater or lesser extent, by considerations of genetic diversity. However, when asked to provide the major overall determinants of the genetic materials used in their breeding programs, breeders generally listed industry requirements and the need to release varieties to ensure the viability of the program ahead of any objectives to broaden genetic diversity.

A common view was that current funding arrangements and industry pressures have been forcing breeders to concentrate more (and more than they felt was desirable) on the short-term goal of releasing new varieties. While they recognised that variety release is ultimately the goal of their program, they argued that more resources should be devoted to exploring wider crosses and more diverse genetic material to ensure that the flow of new improved varieties can continue into the future.

Through information obtained directly from the breeders on perceptions of diversity, breeding priorities, and constraints, we can better understand the importance of the institutional setting in which they operate and its role in shaping the varieties that are ultimately supplied at the farm level. Local and national agricultural policies are clearly major influences on the practical decisions of managing a breeding program, and the analysis in Chapter 4 provides additional detail on the impacts of three broad policy regimes in Australia on diversity levels. It must be left to more rigorous analysis, however, to test for the significance of linkages between changes in the policy environment and changes over time in breeders' perceptions and utilization of diversity.

REFERENCES

Bellon, M.R. 2004. Conceptualizing interventions to support on-farm genetic resource conservation. *World Development* 32: 159-172.

Brennan, J.P., D. Godden, M. Smale, and E. Meng. 1999. Breeder demand and utilisation of wheat genetic resources in Australia. *Plant Varieties and Seeds* 12: 113-127.

Brennan, J.P. and P. Fox. 1998. Impact of CIMMYT varieties on the genetic diversity of wheat in Australia, 1973-1993. *Australian Journal of Agricultural Research* 49: 175-178.

Brennan, J.P. and J.D. Mullen. 2002. Joint funding of agricultural research by producers and government in Australia. In *Agricultural Research Policy in an Era of Privatization*. D. Byerlee and R.G. Echeverria (eds). Wallingford, UK: CABI International.

Brennan, J.P. and K.J. Quade. 2004. *Analysis of the Impact of CIMMYT Research on the Australian Wheat Industry*. Economic Research Report No. 25, NSW Department of Primary Industries, Wagga Wagga.

Day Rubenstein, K. and M. Smale. 2004. International Exchange of Genetic Resources, the Role of Information and Implications for Ownership: The Case of the U.S. National Plant Germplasm System. EPTD Discussion Paper No. 119. International Food Policy Research Institute (IFPRI): Washington, D.C.

Duvick, D.N. 1984. Genetic diversity in major farm crops on the farm and in reserve. *Economic Botany* 38: 161-178.

Reif, J.C., P. Zhang, S. Dreisigacker, M.L. Warburton, M. van Ginkel, D. Hoisington, M. Bohn, and A.E. Melchinger. 2005. Wheat genetic diversity trends during domestication and breeding. *Theoretical and Applied Genetics* 110: 859-864.

Rejesus, R.M., M. Smale, and M. van Ginkel. 1996. Wheat breeders' perspectives on genetic diversity and germplasm use: Findings from an international survey. *Plant Varieties and Seeds* 9: 129-147.

Smale, M. (ed.). 1998. *Farmers, Gene Banks and Crop Breeding: Economic Analyses of Diversity in Wheat, Maize and Rice*. Kluwer Academic Publishers, Boston.

Smale, M. (ed.). 2006. *Valuing Crop Diversity: On-farm Genetic Resources and Economic Change*. Wallingford, UK: CABI Publishing.

Smale, M., E. Meng, J.P. Brennan, and R. Hu. 2003. Determinants of spatial diversity in modern wheat: examples from Australia and China. *Agricultural Economics* 28: 13-26.

Smale, M. and McBride, T. 1996. Understanding global trends in the use of wheat diversity and international flows of wheat genetic resources. *CIMMYT 1995/96 World Wheat Facts and Trends: Understanding Global Trends in the Use of Wheat Diversity and International Flows of Wheat Genetic Resources*. Mexico, D.F.: CIMMYT.

Souza, E., P.N. Fox, D. Byerlee, and B. Skovmand. 1994. Spring wheat diversity in irrigated areas of two developing countries. *Crop Science* 34: 774-783.

Warburton, M.L., J. Crossa, J. Franco, M. Kazi, R. Trethowan, S. Rajaram, W. Pfeiffer, P. Zhang, S. Dreisigacker, and M. van Ginkel. Bringing wild relatives back into the family: recovering genetic diversity in CIMMYT improved wheat germplasm. *Euphytica* 149: 289-301.

Chapter 6

Wheat Diversity Changes in China, 1982-97

R. Hu and E. Meng

ABSTRACT

Changes in production and consumption patterns for wheat in China have combined with changing production constraints to induce new wheat breeding priorities since 1950. These factors together have shaped the selection of new genetic materials by wheat breeders and have in turn induced changes in the level and trends of wheat diversity in China. Using three taxonomies—namely named varieties, morphological characteristics, and pedigree data—we compared measures of spatial diversity across seven key wheat-producing provinces from 1982 to 1997. Diversity based on genealogical and morphological data has generally increased since 1982, though there have been some significant differences between provinces, with Hebei showing a higher and Sichuan showing a lower level of diversity than the other provinces. Diversity based on named varieties has not exhibited precisely the same pattern. Nevertheless, with notable exceptions, diversity at the province and national levels in China has generally been high by international standards, whichever measure is used.

Several recent empirical studies on diversity in major food crops in China, including rice, wheat, maize and soybean, show that widespread adoption of modern varieties has not resulted in declining crop genetic diversity over time, as measured by a range of indicators (Hu et al. 2002; 2000; Wang et al. 2001; Rozelle et al. 2000; Huang et al. 2000). Changes in the structure of the research system and in crop research priorities are likely important contributing factors to changes in the observed level of crop diversity. However, little information is available in the literature on linkages between changes in research objectives and crop diversity.

The overall goal of this chapter is to create a framework for examining the changes in wheat diversity in China. The chapter is organized as follows.

First, we present a general overview of the changes in wheat breeding objectives in China and place observed changes in levels of wheat diversity in China since 1950 in the context of these changes. We then focus on changes in spatial diversity of wheat for different taxonomies and at various levels of aggregation for 1982-97. In the final section, we briefly discuss the implications of the findings.

CHANGES IN BREEDING OBJECTIVES AND THE ADOPTION OF MAJOR PARENTS

Despite there being no significant changes in cultivated wheat area in China during the period from 1949 to the late 1990s, wheat production increased more than 6.5 times in the same period, due primarily to major advances in yield. Although area cultivated in wheat dropped sharply in the late 1990s, due to competition from crops such as maize, yields have maintained a positive trend (Figure 6.1). Increases in both the quantity and quality of inputs used, as well as improvements in production conditions through irrigation, have all contributed to the five-fold increase in wheat yields observed over this time period. Agricultural research focusing on the development and release of modern varieties has also been recognized as a key factor in the yield increase (Fan et al. 2006; Hu 1998). Within this period, 4-6 generations of variety replacement have taken place (He et al. 2001).

Diversity outcomes observed in China are a result of both the supply of new varieties and farm-level adoption decisions. In each year since 1985, more than 300 wheat varieties have been grown by farmers on an area greater than

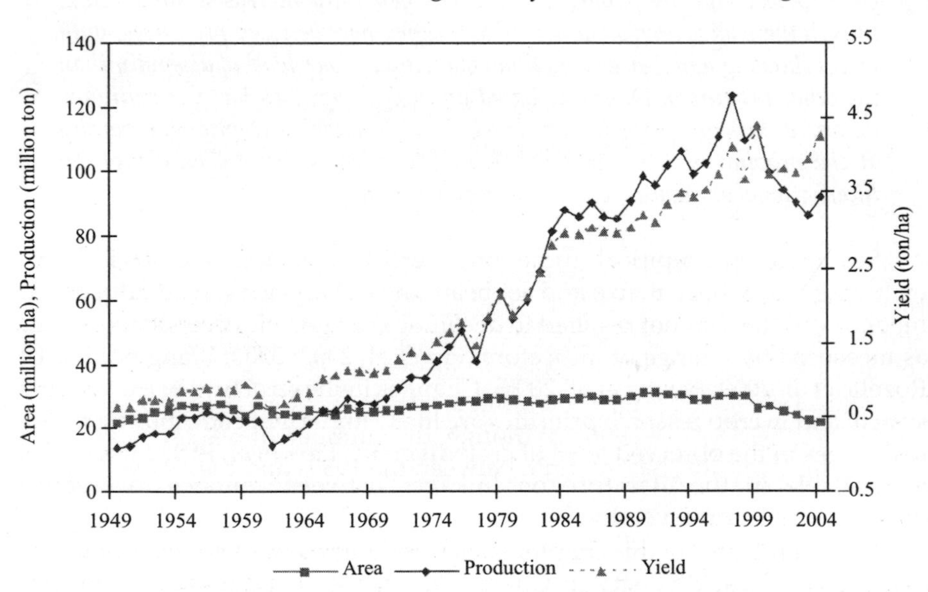

Figure 6.1 Changes in wheat area, yield, and production in China, 1950-2004.

6,667 hectares.[1] The total cultivated area for these 300-plus varieties covers roughly 70% of the total area sown in wheat with the remaining 30% covered by other varieties sown on less than 6,667 hectares each. Figure 6.2 shows the changes in the number of wheat varieties cultivated by farmers in 15 important wheat-producing provinces during 1982-2000. The number of varieties cultivated increased sharply until 1985, after which time only relatively minor changes in total numbers occurred. Winter wheat varieties contributed most to the increase and subsequent variation in wheat variety numbers; relatively small changes have taken place in the number of spring wheat varieties over 1982-2000.

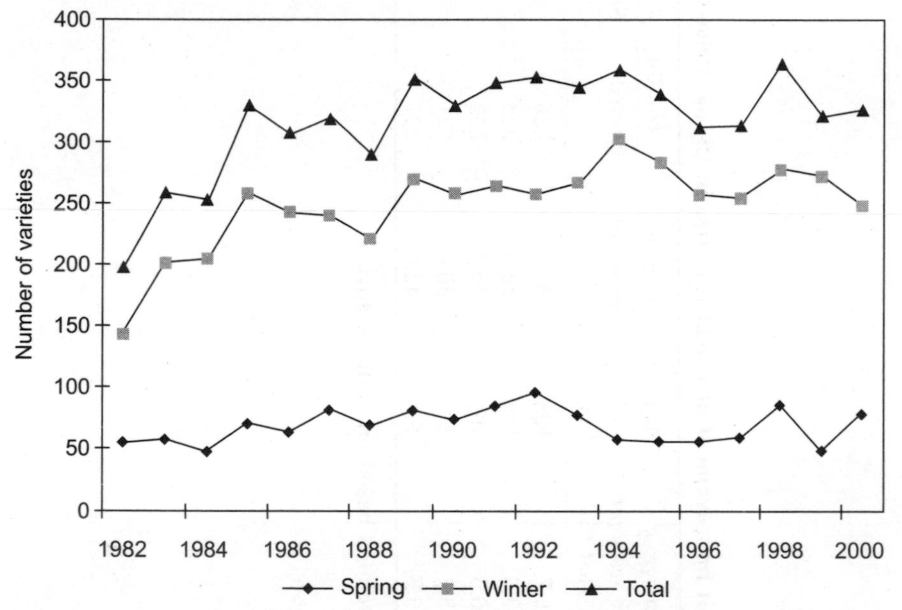

Figure 6.2 Number of varieties cultivated by farmers in China, 1982-2000[2].

The number of varieties cultivated reflects the increase in the available supply of wheat varieties released by breeding institutions in China. There are more than 200 wheat breeding research units in China; these include central, provincial and prefectural levels of agricultural institutes, universities, and a smaller number of private breeding research institutes. More than 5,000 scientists were directly involved in wheat crop research in 2002. As presented in Table 6.1 on page 88, the total number of released wheat varieties doubled from the 1950s to the 1990s, although the number of releases showed a decreasing trend for the first time between the 1980s and the 1990s.

[1]*Data for wheat varieties are collected only for varieties with sown areas exceeding 6,667 hectares or 100,000 mu (1 hectare = 15 mu).*

[2]*Varieties sown on area greater than 6,667 hectares (100,000 mu) per province.*

Table 6.1 Trait improvement of wheat varieties in China, 1950s-90s.

	Number of varieties released	*Variety yield in trials (t/ha)*	*Percentage of varieties resistant to stripe rust (%)*	*Plant height (cm)*	*1000 kernel weight (g)*	*Duration (days)*		*Quality (%)*	
						Winter	*Spring*	*Protein*	*Lysine*
1950s	283	4.1	6.7	108	35.3	226	100	13.6	0.40
1960s	495	5.0	32.8	105	38.6	223	98	11.9	0.39
1970s	540	5.1	60.5	97	40.5	222	98	11.7	0.38
1980s	620	6.0	75.2	90	40.9	221	97	14.0	0.40
1990s	580	7.3	80.7	87	42.0	221	95	14.2	0.41

Source: 1950s-70s: Jin 1983; 1980s-90s: Calculated based on authors' survey.

Table 6.1 also provides information on changes in morphological characteristics for the varieties released by China from the 1950s to the 1990s. Patterns over time in those characteristics clearly reflect significant changes in wheat breeding priorities in China. In accordance with the top priority given to China's food security goals and the key role of cereals such as wheat, yield improvement was the major breeding objective in the early 1950s. At that time most of the wheat varieties cultivated by farmers were traditional local farmer varieties. Average wheat yield was only 0.63 t/ha in 1950 when the government launched a "movement" of participatory variety selection (Jin 1983). The government encouraged farmers, scientists, and officials to make selections from well-performing local varieties, with the main objective of improving wheat yields through dissemination of these relatively higher-yielding varieties. Youzi Mai, Mazhamai, and Pingyuan 50 were some of the local varieties selected and promoted during this time, and the maximum adopted area of each of these varieties was over 600,000 hectares (Jin 1983).

Stripe (yellow) rust also became a major problem in the winter wheat area in China in the early 1950s. Most local varieties were susceptible to the disease, and the associated losses resulted in the addition of stripe rust resistance as a major breeding objective. Local varieties were subsequently replaced by modern varieties bred with disease resistance, including Bima 1, Nongda 183, and Nanda 2419. Bima 1 was the first variety for which the cultivated area reached 6 million hectares. This variety's pedigree includes the variety Quality from Australia and Mazhamai, a local variety from Shaanxi Province. The cultivated area of Nanda 2419, introduced from Italy as Mentana and subsequently renamed, reached 5 million hectares (Jin 1983). More recent sources of stripe rust resistant parents are materials such as Lovrin 10, Predgornaia 2, and Kavkaz, all of which contain the 1BL/1RS rye translocation known to carry specific genes for stripe rust resistance (He et al. 2001). Table 6.1 shows the improvement in resistance to stripe rust for newly released varieties since 1950s. The percentage of varieties resistant to stripe rust increased from close to 7% in the early 1950s to slightly over 80% in the 1990s.

Information on the shifts in breeding objectives since the 1950s is presented in Table 6.2. An important change that took place in the 1960s was the addition of lodging resistance as a breeding objective. Most local, farmer varieties, as well as new releases to that time such as Bima 1 and Nanda 2419, lodged with greater frequency due to increasing farm-level fertilizer use at the end of 1950s. As a result of this breeding objective, plant height declined from 108 cm in 1950s to 87 cm in the 1990s (Table 6.1). The first semi-dwarf parent developed by Chinese scientists was Xiannong 39 in 1964 using the parent Suwon 86 (bearing *Rht1* and *Rht2* genes) and Xinong 6028, a cross between a landrace reselection and a variety introduced from Italy. Following that development, wheat germplasm from several foreign sources, including Italy, the former Soviet Union, and Germany, was introduced into China and used

Table 6.2 Changes of breeders' breeding objectives in China since 1950.

	High yield	*Disease resistance*	*Lodging resistance*	*Short duration*	*High quality*
1950-1954	***				
1955-1959	***	***			
1960s	***	***	***	***	
1970s	***	***	***	***	
1980s	***	***	*	*	**
1990s	***	***	*	*	***
2000s	***	***	*	*	***

Source: Zhuang 2003, Hu 1998, Wu 1990, and authors' survey.

in wheat breeding programs. Some of the most popular parents included Funo, an Italian variety with good yield potential and adaptability; Abbondanza, a widely adaptable Italian variety with resistance to yellow rust; Orofen, a Chilean variety characterized by resistance to stripe and stem rusts and wide adaptability; and St2422/464, an Italian variety with short stature and good adaptability (Jin 1983; Zhuang 2003; He et al. 2001).

Breeding objectives are likely to reflect wider political and social issues. In the case of China, population stress exacerbated the pressing problem of food security from the late 1960s. In continued efforts to ensure adequate grain supplies for the increasing population, the government modified its agricultural production policies and encouraged farmers to produce more grain and other agricultural products by increasing cropping intensity. However, the long duration of the most commonly cultivated wheat varieties posed constraints to the intensification efforts. The selection of shorter duration varieties therefore became an additional breeding objective (Hu 1998). Because many of China's local varieties and other wheat germplasm were early maturing, it was relatively straightforward for Chinese breeders to develop short duration varieties. Table 6.1 shows that the duration of winter and spring varieties released in the 1960s was 2 to 3 days shorter than that of varieties released in the 1950s. The average duration of wheat varieties decreased five days from the 1950s to the 1990s (Table 6.1). Some of the most important parents used in Chinese breeding programs for shortening duration included Youzimai, Mazhamai, Sanyuehuang, Jiangdongmen, and Xiaohongmang (Jin 1983; Zhuang 2003; He et al. 2001). All of these varieties were local varieties singled out during the first yield improvement phase of the early 1950s.

Breeding objectives remained relatively constant in the 1970s; however, due to a grain surplus and corresponding difficulties experienced by farmers in marketing their grain during the 1984 bumper harvest year, attention was increasingly focused on problems with grain quality. Grain quality improvement research projects were initiated in China at that time (Hu 1998), and quality improvement became one of the major breeding objectives. Table 6.1 provides information on the significant increase in the percentage of grain protein and lysine in varieties that have been released since the 1980s. By this time, lodging resistance and short duration traits were routinely screened for, in most breeding programs. They remain breeding objectives, but the emphasis has remained on high yields, disease resistance, and quality (Zhuang 2003).

The widespread use of a key pool of genetic materials associated with desired traits in China's breeding programs and the subsequent adoption of an increasing number of newly released varieties developed from these parents has raised concerns about genetic uniformity. Using information based on coefficients of parentage (COPs) (Cox et al. 1985; Souza et al. 1994), Figure 6.3 shows the area-weighted contribution of selected parental lines used heavily in crossing programs in China and which contributed to China's wheat production during 1982-97. For each replacement generation, some degree of novel genetic variability has been introduced into breeding programs to create new varieties with the desired characteristics, including disease resistance or traits to address other constraints in commercial production (Jin 1983; Hu 1998). Particularly as of the late 1980s, foreign materials have been introduced and dependence on a small core has decreased (Zhuang 2003). A more detailed examination of diversity outcomes in wheat production may shed additional light on linkages between diversity levels and changes in breeding objectives and associated policy priorities.

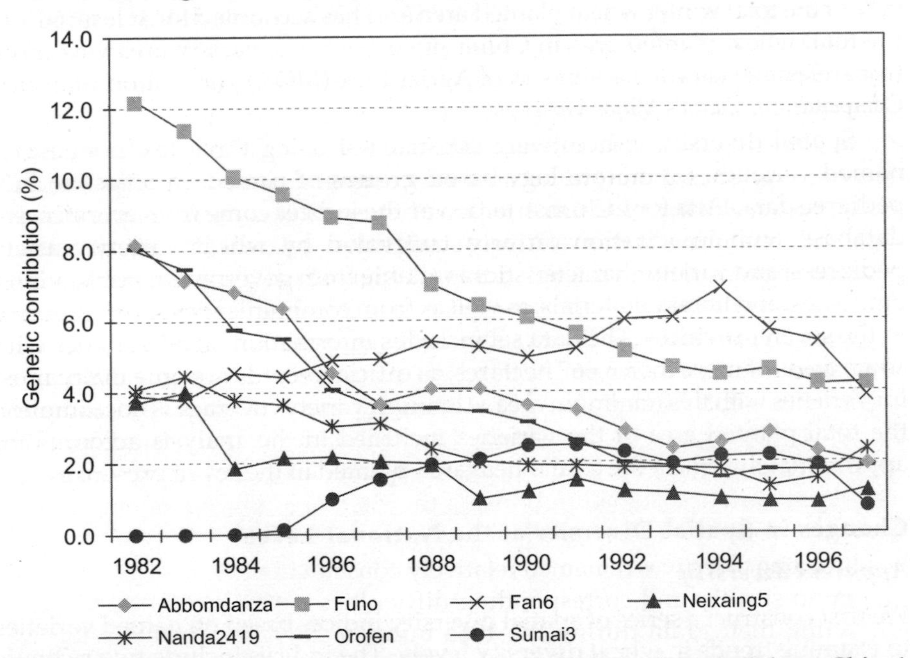

Figure 6.3 Area-weighted contribution of selected genetic materials to China's wheat production, 1982-97.

WHEAT DIVERSITY TRENDS AT PROVINCIAL AND NATIONAL LEVELS

Methodology and Data Sources

Changes in wheat production and consumption patterns, as well as environmental stresses, have induced changes in wheat breeding objectives. These in turn have shaped the selection of new genetic materials for wheat breeding. All of these factors combine to influence the changes in levels and trends of wheat genetic diversity in China. In this section, we examine observed levels of wheat diversity in detail in 1982-97 specifically in the context of breeding objectives and broader policy priorities. In addition to examining diversity trends over time, we focus attention on comparisons across taxonomies for a given diversity concept and comparisons across different diversity concepts using a given taxonomy.

All diversity indices used in this chapter are based on data from the seven major wheat-producing provinces of Shandong, Henan, Hebei, Jiangsu, Anhui, Shanxi, and Sichuan. Shandong, Henan, and Hebei Provinces are located in the Yellow and Huai River Valley winter wheat growing regions; Shanxi is located in the North China winter wheat growing region; Jiangsu and Anhui Provinces are located in the Yangtze River Valley spring/facultative wheat growing region; and Sichuan Province is located in the southwest spring wheat growing region.[3] The area planted to wheat in these seven provinces is approximately 90% of the total winter wheat planted area and has accounted for at least 60% of the total wheat planted area in China since 1982. The variety area was taken from relevant years of the Ministry of Agriculture (MOA) publication, *Statistical Compendium, China's Major Varieties.*

Spatial diversity indices were constructed using three taxonomies: (1) named varieties, (2) morphology-based groups of named varieties, and (3) pedigree data. Data for the construction of the indices come from an extensive database with information on area cultivated by wheat variety, variety pedigrees, and variety characteristics available from government publications, databases, and library materials, as well as from communications with breeders in the seven provinces. The data set includes information on all varieties with sown areas of more than 6,667 hectares, as official records are only maintained for varieties with this minimum area. Although variety coverage is not complete, the total planted area of the varieties included in the analysis accounts for approximately 85% of the total wheat area planted in the seven provinces.

Changes in Spatial Diversity at the National Level

Apparent diversity

We first construct a series of spatial diversity indices based on named varieties to examine trends in wheat diversity levels. The indices include two richness

[3] *Spring and facultative varieties are cultivated during the winter season in both the Yangtze River Valley and in Sichuan Province.*

indices (the number of named varieties and the Margalef index), one measure of inverse dominance (the Berger-Parker index), and one measure of evenness (the Shannon index). All have been defined in Chapter 2. Table 6.3 shows the observed mean, standard deviation, range, and pair-wise correlation among the indices. The two richness indices are highly correlated; however, across the different representations of spatial diversity, there is less correlation in the indices, providing some confirmation of the value of additional information on varietal distribution.

Table 6.3 Descriptive statistics for spatial diversity indices in China, 1982-97 (named varieties).

				Correlation			
Index	*Mean*	*Standard deviation*	*Range*	*# of varieties*	*Margalef*	*Berger-Parker*	*Shannon*
Average # of varieties per province	24.62	8.90	6-49	1.000	0.945	0.463	0.831
Margalef index	1.90	0.75	0.43-3.90	–	1.000	0.411	0.880
Berger-Parker	5.72	2.07	1.58-10.25	–	–	1.000	0.509
Shannon index	2.38	0.46	1.18-3.30	–	–	–	1.000

Figure 6.4 shows changes in the trends of spatial diversity indices using data aggregated across the seven provinces with the index base of 1982 = 1.0. Richness based on named varieties increases steeply during the first half of the 1980s, indicating that diversity defined by varieties in a given area increased significantly during that period. Richness subsequently remained relatively stable until the mid-1990s, when it began to decrease, with a particularly large decrease after 1995, although the decline in the Margalef index is more muted than that of the actual count of varieties. One possible contributing reason for the decrease may be the larger number of varieties with relatively small cultivated areas that make up a larger percentage of total wheat cultivated area in these provinces during the mid-1990s (varieties sown on less than the official benchmark of 6,667 hectares are not reflected in the variety count).

The Shannon evenness index generally exhibits trends similar to those of the two richness indices. It increases to some extent prior to 1985, after which it remains at a steady level with some drop after 1995. However, the two peaks in the Berger-Parker index of inverse dominance in 1987 and 1994 suggest that, while fewer varieties tended to dominate wheat production during the late 1980s, a smaller number of varieties with widespread acceptance were cultivated during the early 1980s and the 1990s. During the late 1990s, the domination of fewer varieties with large cultivated areas re-occurred. The years of peak dominance generally reflect the cultivation of varieties with high yield potential and wide adaptability. For example, Bainong 3217, the variety

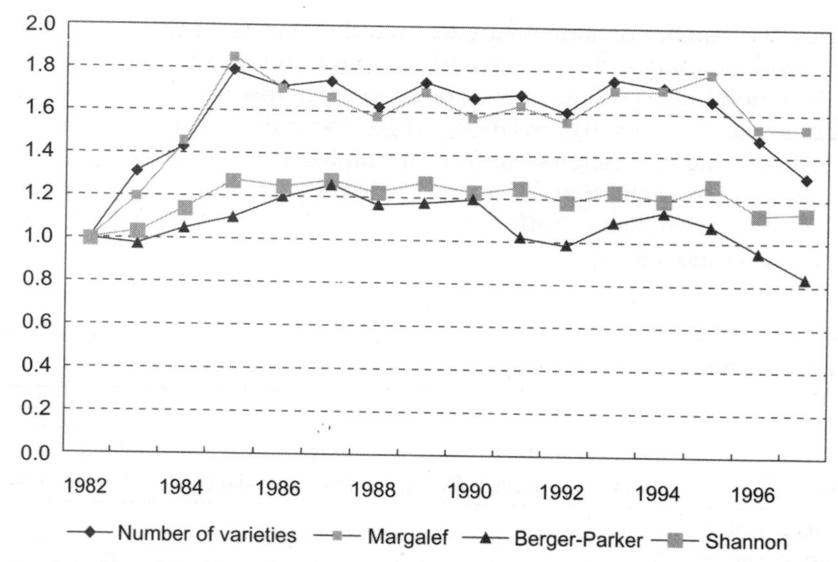

Figure 6.4 Spatial diversity indices for wheat in seven provinces of China, 1982-97 (named varieties).

with the largest cultivated area in China during 1985-1990, performed well not only in land characterized by high soil fertility but also in land characterized by medium soil fertility. The smaller pool of varieties dominating during the mid-latter part of the 1990s generally exhibited high yield potential, better disease resistance and good grain quality. The variety Yangmai 158 was widely adopted in the Yangtze River basin for its excellent disease resistance and high yield potential, while Yumai 18, widely grown in Henan Province, is known for good yield potential and grain quality.

We calculated a similar set of spatial diversity indices at the national level, based on groups formed through a statistical classification methodology using morphological characteristics as described in Chapter 2 (Figure 6.5). Each named variety is associated with one of the groups, and cultivated area for each group calculated accordingly. The characteristics used for the classification were habit, resistance to stem rust, duration, height, and kernel weight at time of release.[4] The mean group values of some of the selected characteristics, as well as mean group yields, are provided in Table 6.4.

In Figure 6.5, the Margalef richness index based on morphological groups, similar to the same index calculated for named varieties, reflects a higher level of diversity relative to the other spatial indices. Richness in morphological groups increased in the early 1990s, both as evenness across groups began decreasing and as one or more groups became increasingly dominant. The lowest point in time in spatial diversity as represented by richness in groups, however, coincided with increasing evenness and decreasing dominance

[4]*Several other traits available for the analysis were not used due to their high level of correlation with the selected traits.*

Table 6.4 Mean characteristics of wheat morphology groups grown in seven major wheat-producing provinces of China from 1982 to 1997*.

Group	*Yield*	*Kernel weight*	*Kernel number*	*Duration*	*Height*
1	355.60	41.68	35.13	242.74	85.86
2	325.51	43.53	45.01	183.32	88.15
3	281.38	38.02	30.00	258.77	102.57
4	396.58	40.53	35.16	234.81	84.46
5	361.22	42.54	35.61	237.25	83.14
6	293.84	37.82	37.28	204.50	100.20
7	336.17	40.41	30.39	250.28	90.19
8	356.50	36.24	43.80	205.00	73.33
9	372.78	39.33	35.76	228.41	87.26
10	302.13	36.46	33.33	96.80	91.17

* Based on trial data at time of release.

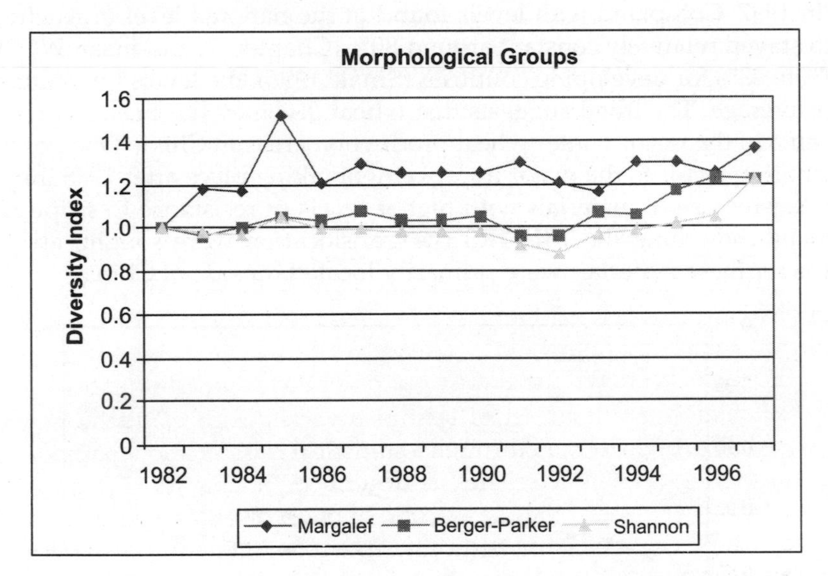

Figure 6.5 Spatial diversity indices for wheat in seven provinces of China, 1982-97 (morphological groups).

among groups. The period between 1993 and 1997 was characterized by generally increasing diversity levels across all indices based on morphological groups. Interestingly, this trend is not reflected in the indices based on named varieties shown in Figure 6.4.

Latent diversity

To examine genetic relationships among the varieties, we also constructed indices of latent diversity using information on variety pedigrees to measure the degree of genetic commonality among ancestors across varieties. For this purpose, we calculate a matrix of pair-wise coefficients of parentage (COPs) and area-weighted coefficients of parentage (WCOPs) for the mix of varieties being grown in a given area in a particular year. As described in Chapter 2, the COP uses genealogies to estimate the genetic relationship between two cultivars based on Mendelian rules of inheritance. The average coefficient of diversity (COD), or one minus the average coefficient of parentage, is a means of reflecting diversity in a set of cultivars grown by farmers. By weighting the COP with area shares to obtain the WCOP, we incorporate aspects of the economic, agroecological, policy, and technical factors determining cultivar diffusion. Changes in these measures over time can indicate the extent to which diversity in cultivated wheat area in farmers' fields is being eroded or enhanced, based on changes in the adoption levels of the varieties being cultivated.

Figure 6.6 shows the changes in COD and WCOD values based on aggregated data across the seven provinces. The values of COD and WCOD increased from 88% and 90%, respectively, in 1982 to levels of approximately 95% in 1997. Compared with levels found at the national level in Australia, which stayed relatively constant around 80% (Chapter 7), and mean WCODs of 79% to 83% for developing countries (Smale 1996), the levels for China are above average. The trend suggests that wheat diversity has been improving over time in the major winter wheat production areas in China. One possible contributing factor to the major improvement taking place after 1988 may be the extensive use of materials with higher levels of resistance to stripe rust. Materials containing the 1BL/1RS rye translocation were sought, and the sources for these materials were primarily located outside of China.

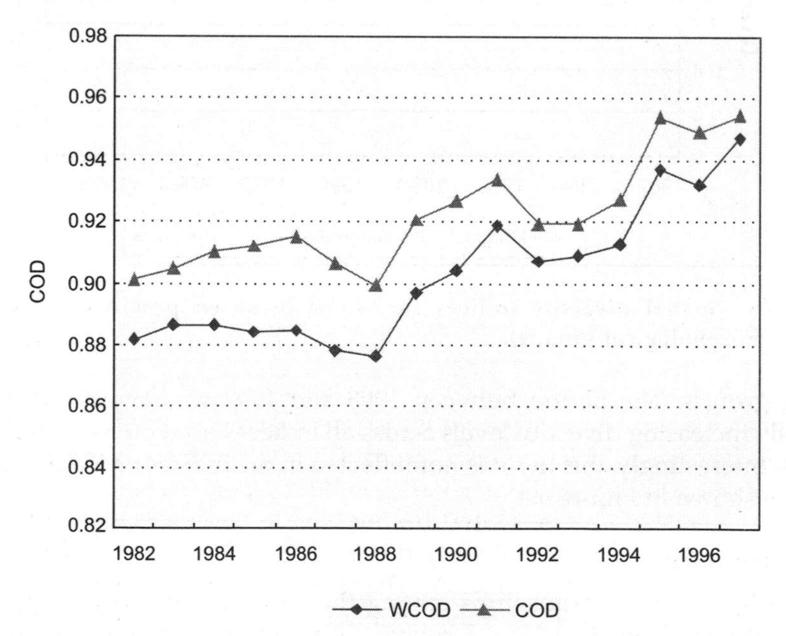

Figure 6.6 Average and area-weighted coefficient of diversity in China, 1982-97.

Changes in Spatial Diversity at the Provincial Level

Apparent diversity

Spatial diversity indices based on named varieties grown in the seven major wheat-producing provinces of China between 1982 and 1997 are presented in Figure 6.7. A comparison across the richness, dominance, and evenness indices calculated using the data set of named varieties shows, in general, wheat areas in Hebei and Henan to be the most diverse throughout the study period, while Sichuan and Jiangsu wheat regions are most often among the least diverse across the diversity indices. In terms of richness (Figures 6.7 (a) and (b)), farmers in Hebei and Henan generally cultivated a larger pool of varieties, whereas farmers in Sichuan cultivated the smallest pool of varieties during most of 1982-97.

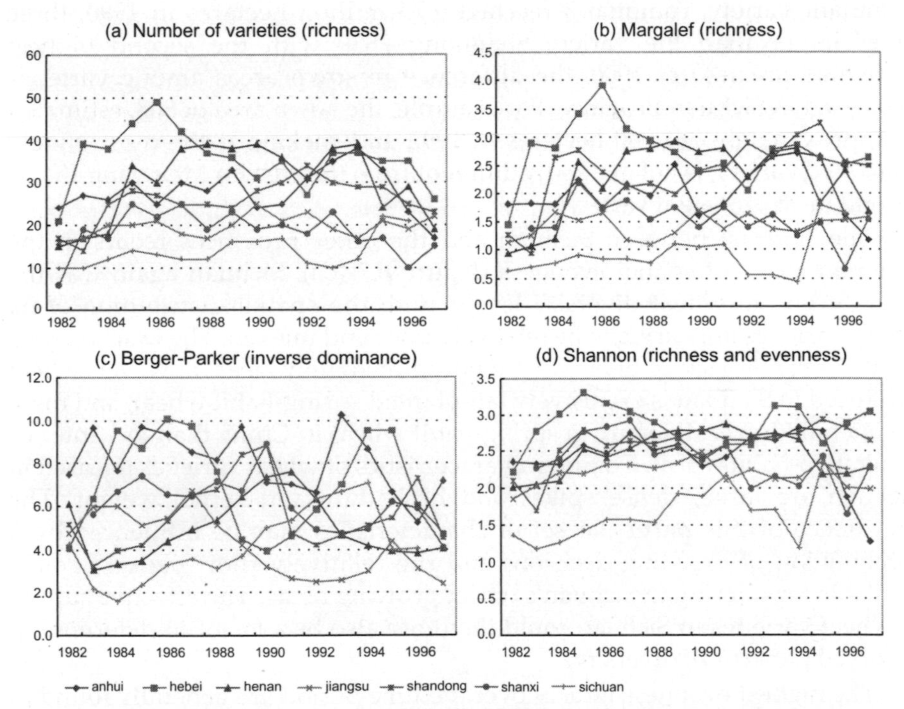

Figure 6.7 Spatial diversity indices by province, 1982-97 (named varieties).

A peak in the Berger-Parker index expresses low levels of dominance in the cultivated pool of varieties (Figure 6.7 (c)). With no variety clearly dominant over others, the Berger-Parker index is likely to be high, since the maximum area share represented by any single variety is relatively low. Inadequate seed supplies relative to demand could also pose constraints to the area planted to certain varieties. The inverse dominance index for each province shows a cyclical trend coinciding with the emergence and disappearance of popular varieties, a trend that is evident in all provinces over the time period. A wheat variety may disappear because it is replaced by new varieties in the pool or because its seed sources gradually diminish, or some combination of both. Inadequate seed supplies relative to demand may also constrain the area planted to popular varieties. No province emerged as clearly superior to another province, although the inverse dominance index reached its lowest levels (associated with the greatest dominance by a single variety) in Sichuan and Jiangsu. Particularly prior to 1990 in Sichuan Province, the difference in sown areas between dominant varieties and other varieties was bigger. For example, the sown area of a dominant variety, Mianyang 11, reached 1.4 million hectares in 1984, which was 12 times larger than the variety Mianyang 15 with the second largest area sown. The same situation appears in Jiangsu Province after 1989. The sown area of the dominant variety Yangmai 5 reached 0.75 million hectares in 1990, three times larger than the variety Shannong 7859 with the second highest cultivated area. After 1990, the difference in sown areas among varieties decreased in Sichuan Province. For example, the sown area of 808, estimated at approximately 361,000 hectares in 1992 and making it the most widely cultivated variety, is not markedly different from the area for Mianyang 15, the second most prevalent variety.

The contrast between Sichuan and the other provinces recurs in the Shannon index of spatial evenness (Figure 6.7 (d)). Sichuan again was the least "even" in wheat diversity levels, with the spatial distribution of its wheat varieties appearing to be relatively poor and uneven. The exact reasons for this relative lack of diversity are not immediately clear; however, wheat produced in Sichuan is exclusively fall-planted spring-habit wheat, and there is generally lower diversity in spring-habit wheat in China than in winter or facultative-habit wheat. Common characteristics of wheat varieties released in Sichuan are large, dense spikes and high thousand-kernel weight. The prevalence of this particular set of characteristics may be influenced by a combination of breeding decisions and the relatively short period of time available for tillering in Sichuan's wheat growing areas. The effective supply of wheat varieties in Sichuan could therefore also be a factor in determining observed patterns of diversity.

The highest evenness indices over the time period are generally found in Hebei Province. A possible explanation for the relative evenness among wheat varieties in Hebei may lie in its agro-ecological suitability for bread wheat varieties of all three growth habits; winter, spring, and facultative. There is similar evenness among wheat varieties and agro-climatic diversity in Shanxi Province, where all three growth habits are also present, but to a lesser extent than in Hebei.

A similar set of spatial diversity indices was calculated based on groups classified for each province using morphological characteristics of varieties cultivated in the province (Figure 6.8).

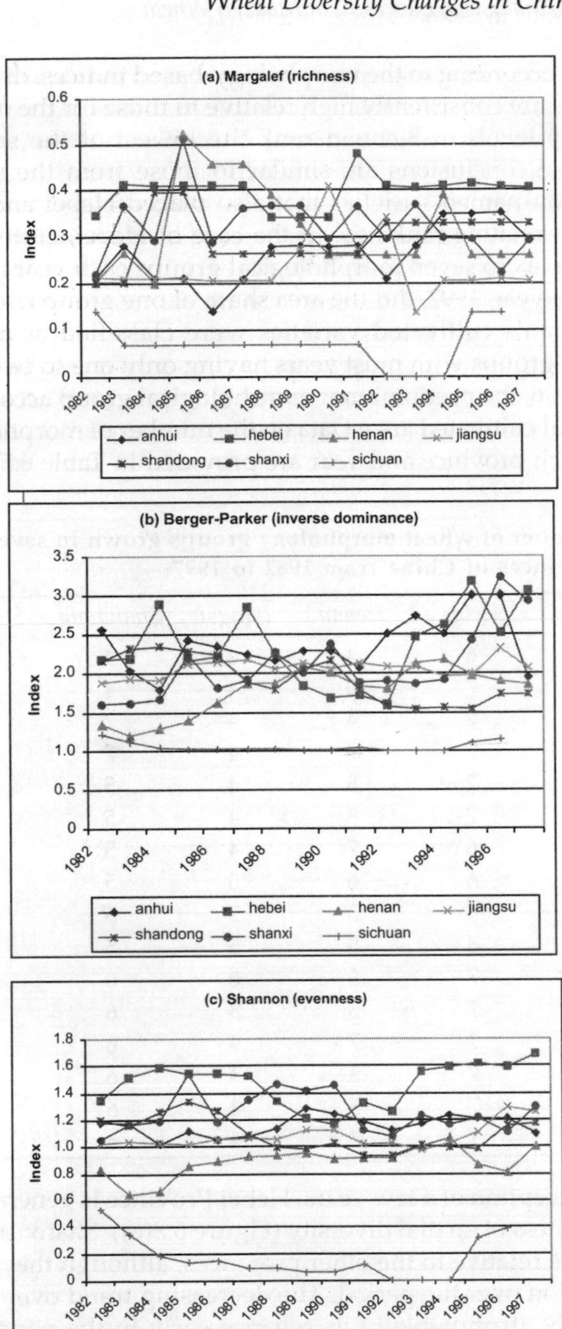

Figure 6.8 Spatial diversity indices by province, 1982-97 (morphological groups).

In general, according to the morphology-based indices, diversity levels in Hebei Province are consistently high relative to those for the other provinces, while diversity levels in Sichuan rank the lowest of the seven provinces analyzed. These conclusions are similar to those from the set of diversity indices based on named varieties that also placed Hebei and Sichuan high and low in the relative rankings. In the case of Hebei, cultivated area was divided among six to seven morphological groups each year in the province, and only in one year, 1992, did the area share of one group rise above 60%. In contrast, Sichuan's cultivated varieties were classified at most into three morphological groups with most years having only one to two groups. In all years except 1996, the predominant morphological group accounted for more than 90% of total cultivated area. Data on the number of morphological groups classified in each province and year are provided in Table 6.5.

Table 6.5 Number of wheat morphology groups grown in seven major wheat-producing provinces of China from 1982 to 1997.

Year	*Anhui*	*Hebei*	*Henan*	*Jiangsu*	*Shandong*	*Shanxi*	*Sichuan*
1982	4	6	4	4	4	4	3
1983	5	7	5	4	4	6	2
1984	4	7	4	4	7	5	2
1985	4	7	9	4	7	8	2
1986	3	7	8	4	5	5	2
1987	4	7	8	4	5	6	2
1988	5	6	7	4	5	6	2
1989	5	6	6	5	5	6	2
1990	5	6	5	7	5	5	2
1991	5	8	5	5	5	6	2
1992	4	7	5	6	6	5	1
1993	5	7	5	3	6	6	1
1994	6	7	7	4	6	5	1
1995	6	7	5	4	6	5	3
1996	6	7	5	4	6	4	3
1997	5	7	5	4	6	5	na

With the exception of a few years, Hebei Province is generally the highest in terms of richness of spatial diversity (Figure 6.8(a)). Shanxi and Henan also have high levels relative to the other provinces, although they exhibit higher levels of variation over the period. The decreasing trend over time in Henan may be partially attributable to its reliance early in the period on varieties from other provinces. This reliance decreased as Henan began cultivating a smaller set of its own released varieties. The levels in Shandong Province, while lower, remain relatively constant throughout the period. In the inverse dominance index based on morphological groups (Figure 6.8(b)), a cyclical pattern similar to the one observed for named varieties again emerges. Hebei and Anhui have relatively higher overall levels of diversity according to this index, but have high levels of variation as well. In Henan Province, morphological groups become less dominant over the time period, and the cyclical nature of shifting groups is also not as evident. Finally, in terms of evenness of morphological groups, Hebei and Shanxi clearly rank at the top of the set of provinces examined. Morphological groups in Henan and Sichuan Provinces appear to be the least evenly distributed. Again, for Henan, the dominance of specific morphological groups and the lack of evenness across groups, relative to other provinces, are likely related to a focus on a narrower set of preferred characteristics.

Latent diversity

Coefficients of diversity (CODs) and weighted coefficients of diversity (WCODs) are calculated for each of the seven provinces to reflect changes in diversity at the province level (Figure 6.9).

As with the apparent spatial indices based on both named varieties and morphological groups, Sichuan Province stands out from the other six provinces with much lower levels of diversity and reinforces information from the spatial diversity indices that diversity levels are lower in Sichuan than in other provinces. A possible explanation for these findings may be related to Sichuan's unique agroecological conditions and the province-specific germplasm commonly used by breeding programs. Relative to other wheat agroecological zones, there is only a small temperature variation in Sichuan during the entire wheat cropping cycle. Wheat duration is approximately 180-190 days in the province, a much shorter period than other winter wheat producing regions, and longer than spring wheat producing regions. The varieties grown in the province are primarily spring types that are planted in the fall. During the colder winter period the plants are at the tillering stage and require some degree of cold hardiness, but not vernalization as with true winter wheat.

In contrast to Sichuan Province, the other six provinces exhibit much higher COD and WCOD values during 1982-1997. Almost all values are greater than 85%, which suggests that wheat diversity as defined by diverse use of breeding materials remained at a high level in most of the winter wheat production regions in China. Shanxi Province exhibits the highest level of diversity thus defined, particularly in the first half of the period, which is interesting in light of the fact that its level of diversity based on named varieties did not stand out amongst the seven provinces. In contrast, the levels of evenness and richness in Shanxi Province's diversity based on morphological groups ranked among the highest of the seven provinces. Both CODs and WCODs in Shandong, Hebei, and Jiangsu Provinces increase sufficiently during the first half of the period to approach and/or equal the level of Shanxi Province by 1991.

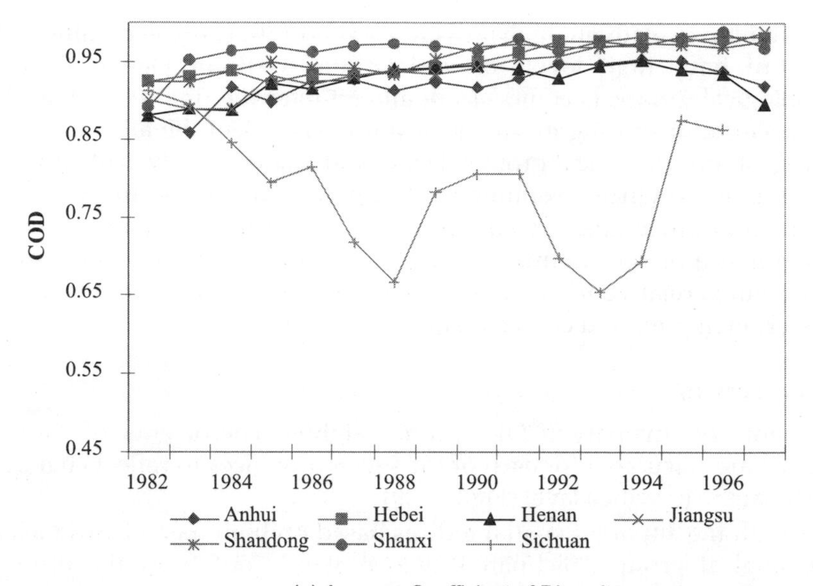

(a) Average Coefficient of Diversity

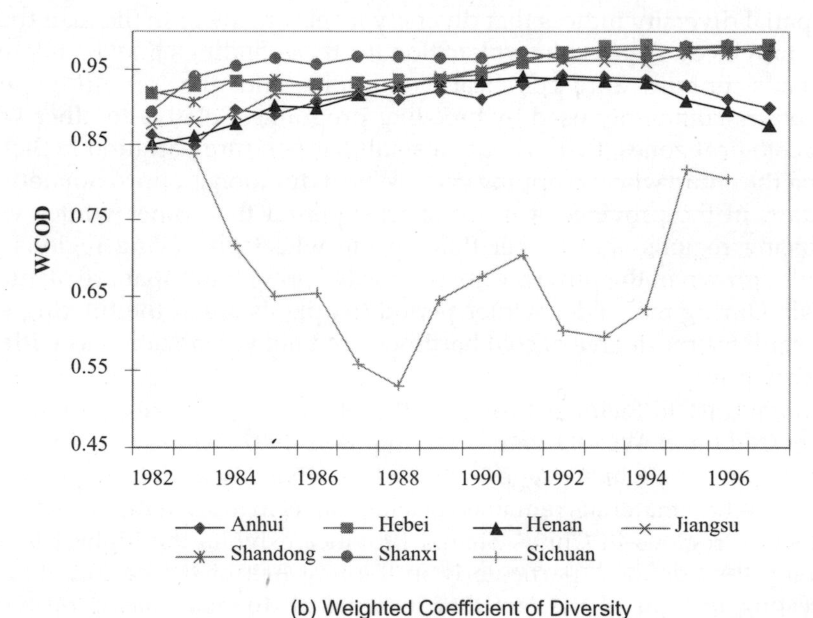

(b) Weighted Coefficient of Diversity

Figure 6.9 Average and area-weighted coefficient of diversity in seven provinces of China, 1982-97.

Overall increasing trends during the 1980s indicate that diversity in breeding materials expanded during this period. One possible reason for this increase is the focused effort of many breeders in the 1980s to broaden the range of sources for breeding materials, particularly materials from outside China, for use as parents in their crossing programs.

COD and WCOD values are generally lower in Anhui and Henan Provinces during the entire period from 1982 to 1997. Moreover, although it shows up slightly earlier in the WCOD, values for both indices begin to trend down in the early 1990s, suggesting decreases in wheat diversity levels. One possible contributing factor was the success of two varieties, Yumai 18 (released in 1990) and Yumai 21 (released in 1992), both of which were developed using the parent Yanshi 4. These varieties were adopted extensively in the two provinces (e.g., approximately 50% of the total area planted in wheat in Henan) in 1996 and 1997.

Discussion and Implications

Comparisons across taxonomies

The implications of using different taxonomies to define crop populations are revealed when indices with identical construction, but using populations differently classified by cultivar name and by morphological characteristics, are compared. In the aggregate data for apparent spatial diversity, variation over time evident in diversity indices based on named varieties is somewhat dampened in the morphological classification. Both taxonomies indicate minimal changes during the 1980s; however, trends of increasing diversity beginning in 1992 and 1993 reflected in all indices based on morphological groups are not reflected by the same indices using named varieties. These indices, in fact, indicate declining trends in diversity beginning in the early-mid 1990s. Latent diversity indices based on genealogies appear to concur that relatively minimal changes in diversity levels took place in the 1980s, but reflect increasing levels of diversity beginning in the 1990s, thus reinforcing the trend reflected in the diversity indices based on morphological groups.

The consistency between genealogy-based diversity indices and morphological group-based indices could be explained to some degree if the traits used for classification were qualitative traits determined by one or few genes that could be associated with a small pool of parents. Of the variety traits used to form the morphological groups, however, only habit and height are generally associated with one or few genes. Furthermore, even though there are a number of dwarfing genes, other underlying genes ensure some variance within a range. Duration and kernel weight are both quantitative traits determined by multiple genes, and stem rust resistance can be conveyed by either a major gene or a group of minor genes.

In the provincial analysis, relative rankings across provinces using three different taxonomies confirmed the diversity of wheat in Hebei Province and the lack of diversity in Sichuan Province, relative to the other provinces. However, there were very few situations in which all taxonomies agreed on trends over time. Diversity indices based on genealogies continued to trend up gradually in certain provinces (e.g., Hebei, Jiangsu, Shanxi, and Shandong) even as many of the indices based on named varieties and morphological groups showed much more variation during the same period. The information provided from the analysis of spatial and genetic diversity show some differences. These differences reflect the importance of assessing latent diversity based on pedigrees in addition to apparent diversity indices based on named varieties and morphological characteristics, where data are available.

Linkages between national and provincial level

At the national level, the COD and WCOD are lower than those found at provincial levels, with the exception of Sichuan Province, indicating that variation in genealogical backgrounds of cultivated varieties are higher in individual provinces than at the national level. However, the national levels are high relative to other countries (Smale 1996) and indicate that the diversity at the province level benefits the country as a whole. Moreover, the national level COD and WCOD maintains an overall increasing trend during the time period examined, despite the performance of three individual provinces, particularly towards the end of the time period. The COD and WCOD at the national level also increase at a much faster rate than those at the provincial level.

Interestingly, these results appear to be the opposite of what was observed in Australia, where lower levels of measured diversity resulted when smaller regions were analyzed, regardless of the measure used (Chapter 7). The higher likelihood of similarities among environments and existing farming systems was hypothesized as the likely rationale in that situation, and larger regions were likely to include a wider range of production environments and farming systems. In China's case, diversity among environments and materials used in breeding programs at the provincial level appears to be much more prevalent. However, as evident in Sichuan, aggregate analysis at the national level can nevertheless fail to reflect changes taking place at a more disaggregated level.

Overall, the analysis of changes in wheat diversity in China for the period 1982 to 1997 reveals a number of trends:

- Spatial diversity based on the number and distribution of groups determined by morphological traits and on genealogies, particularly in winter wheat, has generally improved since 1982. The improvement mainly took place after 1988.
- Significant differences in levels of wheat diversity are observed among different provinces. All taxonomies used found wheat diversity to be higher in Hebei Province and lower in Sichuan Province.
- Observed levels of apparent spatial diversity based on named varieties and latent spatial diversity are not entirely consistent. Apparent spatial diversity measures based on groups classified by morphological characteristics, most of which are determined by single genes, were much more consistent with a pedigree-based measure of diversity.

Changes in breeding objectives have played a major role in the observed levels and changes of wheat diversity in China. It would be useful to test hypotheses that link some of these shifts in breeding objectives to developments related to socio-economic factors, government policies, and production and consumption factors including population stress, economic growth, soil fertility improvement and changes in abiotic stresses.

REFERENCES

Cox, T.S., Y.T. Kiang, M.B. Gorman, and D.M. Rodgers. 1985. Relationship between coefficient of parentage and genetic similarity indices in the soybean. *Crop Science* 25: 529-532.

Fan, S., K. Qian, and X. Zhang. 2006. China: An unfinished reform agenda. In *Agricultural R&D in the Developing World: Too Little, Too Late*? P.G. Pardey, J.M. Alston and R.R. Piggott (eds). Washington, D.C.: International Food Policy Research Institute.

He, Z.H., S. Rajaram, Z.Y. Xin, and G.Z. Huang (eds.). 2001. *A History of Wheat Breeding in China.* Mexico, D.F.: CIMMYT.

Hu, R. 1998. *Seed Industry in China*. China's Science Press. 1998.

Hu, R. 1999. Agricultural R&D Policy and Genetic Diversity in China. Research Report, Center for Chinese Agricultural Policy, Chinese Academy of Sciences.

Hu, R., J. Huang, and S. Rozelle. 2002. Genetic uniformity and its impacts on wheat yield in China. *Scientia Agricultura Sinica.* 35(12): 1442-1449.

Hu, R., J. Huang, S. Jin, and S. Rozelle. 2000. Assessing the contribution of research system and CG genetic materials to the total factor productivity of rice in China. *Journal of Rural Development* 23(2000): 33-70.

Huang, J. and S. Rozelle. 2000. Agricultural Policy, Wheat Economy and Genetic Diversity in China. Working Paper WP-00-E8, Center for Chinese Agricultural Policy, Chinese Academy of Sciences.

Jin, S. (ed.). 1983. *Chinese Wheat Improvement and Pedigree Analysis*. China Agricultural Press.

MOA (Ministry of Agriculture). 1982-2000. *Statistical Compendium, China's Major Varieties.* Unpublished statistical volume. Ministry of Agriculture, Beijing, China.

Rozelle, S., E. Meng, S. Jin, R. Hu, and J. Huang. 2000. Genetic Diversity and Total Factor Productivity: the Case of Wheat in China. Working Paper WP-00-E9, Center for Chinese Agricultural Policy, Chinese Academy of Sciences.

Smale, M. 1996. Understanding Global Trends in the Use of Wheat Diversity and International Flows of Wheat Genetic Resources. Economics Working Paper 96-02. Mexico, D.F.: International Maize and Wheat Improvement Center (CIMMYT).

Souza, E., P. Fox, D. Byerlee, and B. Skovmand. 1994. Spring wheat diversity in irrigated areas of two developing countries. *Crop Science* 34: 774-783.

Wang, Li, Ruifa Hu, Jikun Huang, and Scott Rozelle. 2001. Soybean Genetic Diversity and Production in China. *Scientia Agricultura Sinica* 34(6): 604-609.

Wu, Z. 1990. *Wheat Breeding*. China Agricultural Press, Beijing, China.

Zhuang, Q. 2003. *Chinese Wheat Improvement and Pedigree Analysis*. China Agricultural Press, Beijing, China.

Chapter 7

Wheat Diversity Changes in Australia, 1965-97

J.P. Brennan, A.B. Bialowas, and D. Godden

ABSTRACT

Changes in wheat diversity in Australia between 1965 and 1997 are examined at three levels: national, state, and shires within the state of New South Wales (NSW). Two measures of apparent spatial diversity, richness and evenness of named varieties, as well as the latent spatial diversity based on variety pedigrees, are assessed at each level. At the national level, richness and evenness measures indicate that diversity increased during the period. At the state level, three states had relatively steady diversity, and two had increasing levels. At the shire level, apparent diversity was considerably lower than at the state or national level, and showed some signs of marked decline in some shires, although different parts of New South Wales demonstrated different trends. A different picture emerged from the latent diversity measure based on pedigree data, where diversity declined in later periods in all shires and most states, while remaining high at the national level. This was consistent with an increase in the number of varieties grown, but also where the varieties are relatively closely related genetically. The implication of the findings for policies to address diversity appropriately could be significant.

In Chapter 4, the main policy influences on changes in genetic diversity in Australia were found to be the institutional structures of the wheat industry, the marketing system, technological developments, and the government enacted Plant Breeders' Rights. With that policy background in mind, in this chapter we examine the changes in wheat diversity in Australia from 1965 to 1997. Specifically, we focus on increasingly disaggregated measures of spatial and genetic diversity beginning at the national level, then the state level, and finally the shire (or local) level within the state of New South Wales (NSW).

CHANGES IN APPARENT SPATIAL DIVERSITY AT STATE AND NATIONAL LEVELS

For all three levels of analysis, we focus on two measures of spatial diversity: the Margalef index of varietal richness and the Shannon index of varietal evenness (Chapter 2). Both indices are calculated from data on cultivated area for named varieties. The Margalef index measures the relative richness of the set of varieties in relation to the area sown (the higher the index, the greater the relative richness of the set of varieties grown), while the Shannon index combines information on the richness of the varieties grown with a measure of their relative abundance (the higher the index, the richer and more even the spread of the set of varieties being grown).

Wheat production data in recent years provides an indication of the scale of the industry at each level of disaggregation. Average production through the 1990s for Australia was 17 million tons, while for the main wheat-growing states, average production ranged from 1 to 7 million tons, broadly similar in scale to the major wheat-growing provinces in China (Chapter 6).

The Margalef index at the national level during the period between 1965 and 1997 is shown in Figure 7.1. Apart from brief declines in the late 1960s and early 1980s, a substantial upward trend in varietal richness, from an approximate value of 3.0 in the 1960s to around 8.0 in the 1990s, can be observed throughout most of the period in the index. The period from 1980 until the mid-1990s demonstrates a pattern of steadily increasing varietal richness. However, beginning about 1993, a decline in richness at the national level becomes very evident.

The move toward more differentiated varietal marketing from around 1971 (Chapter 4) and technological developments throughout the period have contributed to the upward trend in the Margalef index until the early 1990s.

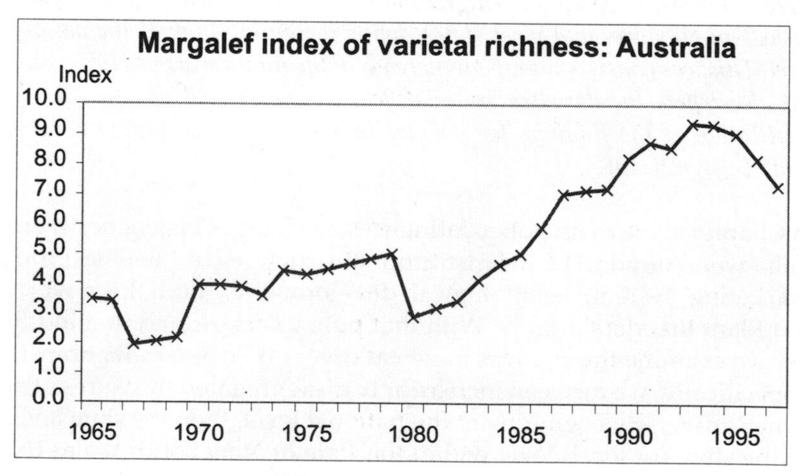

Figure 7.1 Margalef index of varietal richness: national level.

The downward trend in the more recent period may be the result of the more commercialized plant breeding and the increased importance of crops other than wheat in providing an alternative means to diversity on wheat farms.

Changes in varietal richness, as measured by the Margalef index, in each of the main Australian wheat-growing states between 1965 and 1997 are shown in Figure 7.2. Farming technologies are similar across all states, although there are some differences in the farming systems. In the northern area of Queensland and northern New South Wales, summer rainfall is dominant and mainly high-protein hard wheats are produced in cropping systems. In the rest of Australia, with winter rainfall dominant, wheat is generally produced in rotation with pastures. These different systems have led to some differences in the experiences of the states with varietal diversity, although the trend exhibited by the state-level indices indicates a noticeable increase in varietal richness in each state beginning in the early to mid-1980s. The increase was lowest in Queensland, but appears to be similar in each of the other four states. Prior to 1980, most states showed very little change in levels of varietal richness. However, since the mid-1990s a small decline has been observed in all states except Western Australia.

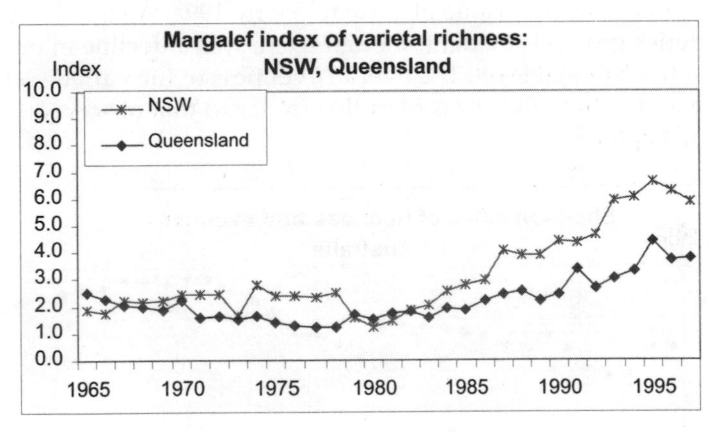

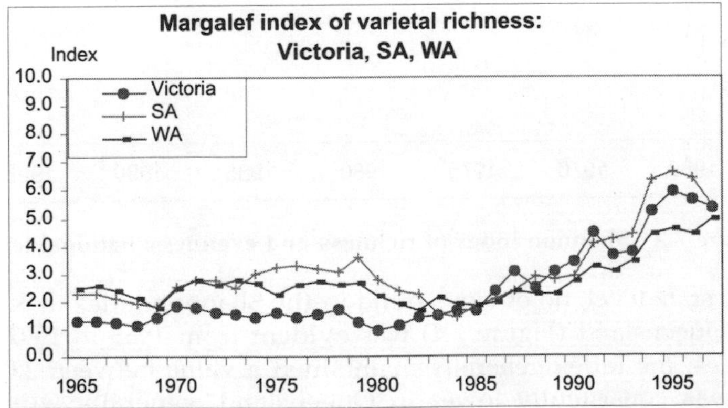

Figure 7.2 Margalef index of varietal richness: state levels.

The movement over time at the national level is therefore not paralleled at the more disaggregated state level. The richness at the national level increased until 1980, despite very little variation in richness of the variety mix within each state. Contributing to this phenomenon is the fact that more varieties were released for specialized production environments within each state during that period, so that while the number of varieties in each state did not change markedly, the richness at the national level was increasing.

However, since approximately 1980, the upward trend in varietal richness at the national level has been matched by an increasingly rich mix of varieties in each state. The data thus highlight the existing practice in that period for each state to develop its own individual variety sets through its own breeding program, but with fewer and fewer varieties cultivated across more than one state. This trend is consistent with the move since the early 1980s away from broadly adapted varieties and toward more specific regional adaptation. It also reflects a general increase in the number of varieties cultivated in each state.

At the national level, the Shannon evenness index (Figure 7.3) increased from a value of 2.5 in 1965 to 3.5 in the late 1980s, but subsequently showed a steady decline back to a value of close to 3.0 by 1997. Again, the increasing mix of varieties grown is evident, though there was a decline in evenness in the 1990s at the national level. The overall evenness of the varieties cultivated at the state and national levels is broadly similar to that observed in China's provinces (Chapter 6).

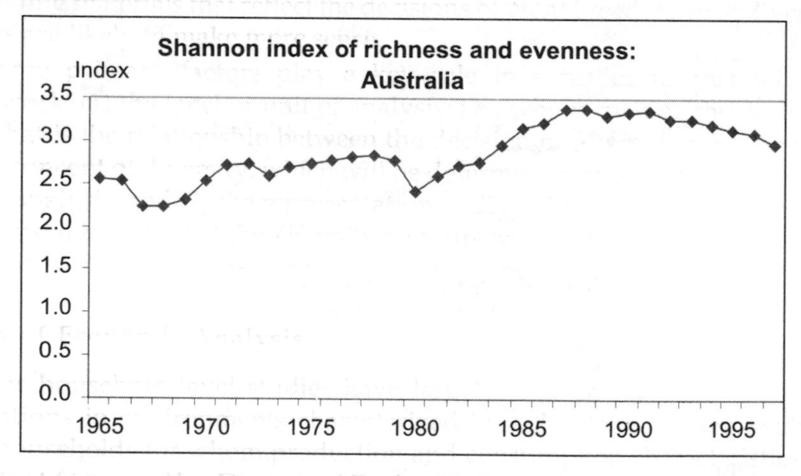

Figure 7.3 Shannon index of richness and evenness: national level.

At the state level, no obvious trend in the Shannon index in New South Wales or Queensland (Figure 7.4) was evident from 1965 to 1997. In New South Wales, the index generally maintained a value between 2.0 and 2.5, while it was consistently lower in Queensland, generally with a value between 1.5 and 2.0. In contrast, Shannon index values in Victoria, South

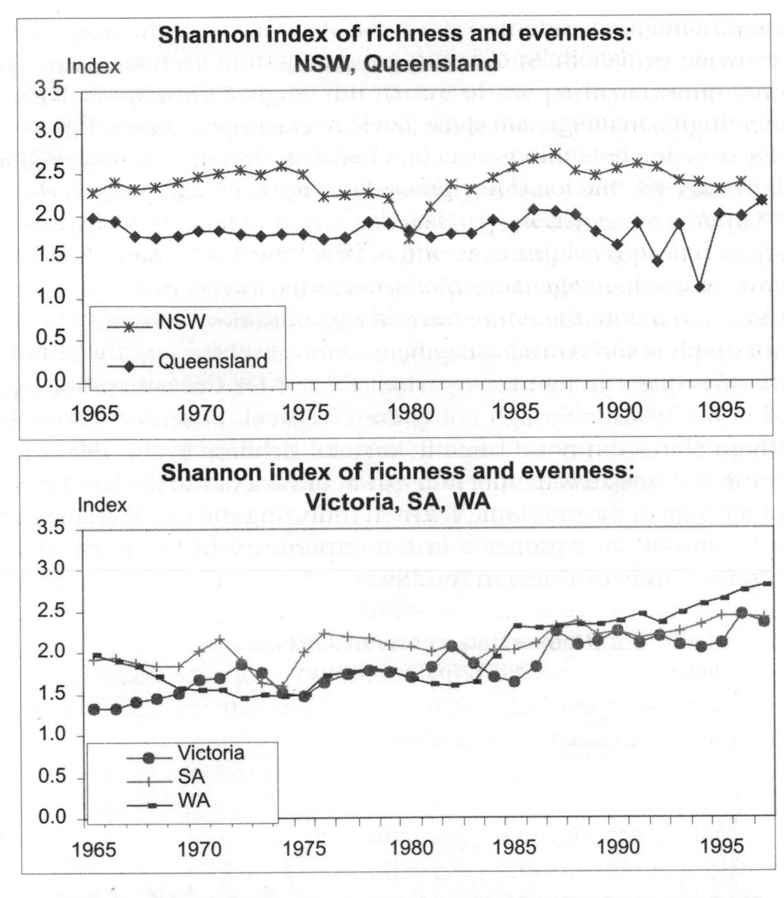

Figure 7.4 Shannon index of richness and evenness: state level.

Australia, and Western Australia all showed an overall increase over the period, rising from between 1.5 and 2.0 in the 1960s to around 2.5 by 1997 in each case. The generally higher level of the Shannon index in New South Wales likely reflects the greater diversity of wheat-growing environments in the state, ranging from areas of winter-dominant rainfall (similar to Victoria, South Australia, and Western Australia) in the south and areas of summer-dominant rainfall (similar to Queensland) in the north. The upward trend in the three southern states reflects a relatively more even spread of varieties over the period since 1965.

Changes in Apparent Spatial Diversity at Shire Level in New South Wales

The changes in the diversity indices for each of eight selected shires categorized by geographical location (Brennan and Bialowas 2001) are now considered. With an average wheat area of 60,000 hectares, the shires are much smaller in area than either the Australian states or the major Chinese wheat-growing provinces. Since environments within each shire are smaller and more uniform than across each state, the range of varieties is likely to be correspondingly smaller at the shire level. Accordingly, varietal diversity is also likely to be lower at the shire level. Very little variation is observed in the Margalef index for the southern shires between 1965 and the early 1980s (Figure 7.5). A rapid increase in richness occurred in the late 1980s, but it was followed by a strong decline during the 1990s. By 1997, the richness of the variety mix in southern New South Wales was the lowest in the past 32 years, largely associated with the dominance of two varieties, Janz and Dollarbird. The trend graph is similar in the northern shires until the late 1980s, but very different subsequent to that time period. Except for Coonabarabran, which declined in the 1990s (although not to the low levels observed in the 1970s), the northern shires did not decline in varietal richness in the 1990s. In fact, most northern shires grew a more diverse set of varieties in the late 1990s than at any other time in the previous 32 years, reflecting the fact that, perhaps by chance, no individual varieties had the superiority in the north that was evident in the southern shires in the 1990s.

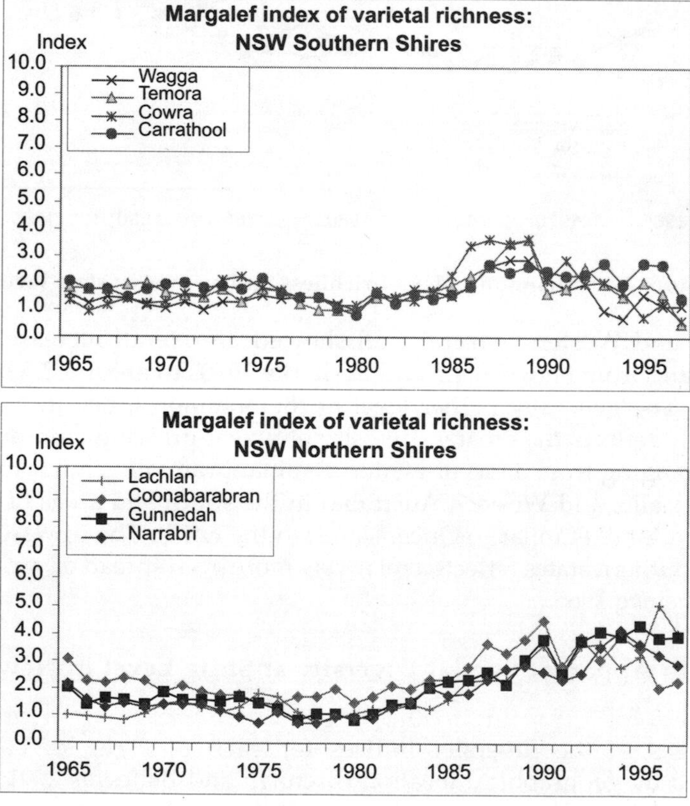

Figure 7.5 Margalef index of varietal richness: New South Wales shire level.

The value of the Shannon index fluctuated in the southern shires, but in general the average value ranged from 1.5 to 1.8 in the 1960s and 1970s. During the 1980s, the average of the index increased in all shires (Figure 7.6). However, there has been a consistent sharp decline in richness and evenness of the variety mix in the 1990s in the southern shires, again, attributable to the dominance of the varieties Janz and Dollarbird in those shires. In contrast, the general level of the richness and evenness index in three of the northern shires increased in the 1990s with the average of the Shannon index increasing to 2.0 or higher and reflecting a move toward a larger and more diverse set of varieties being grown. Only in Coonabarabran, the most western of the northern shires, did a decline take place during the 1990s, though it was not as significant as that taking place in the southern shires. These differences between shires are likely to be the result of largely unplanned events in the development and area dominance of varieties, since the government policy regimes were for the most part indistinguishable across all shires.

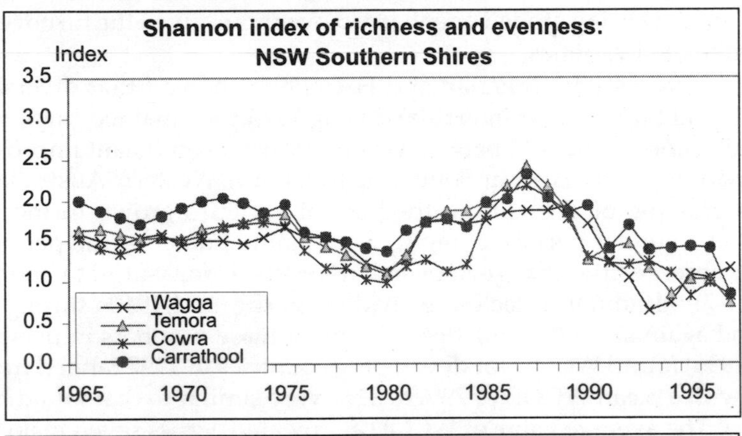

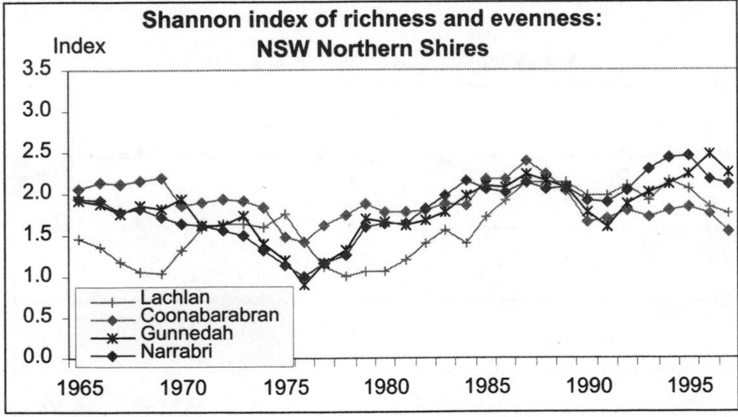

Figure 7.6 Shannon index of richness and evenness: New South Wales shire level.

Changes in Latent Diversity at State and National Levels

The underlying basis for the measures of spatial diversity discussed previously is the set of varieties cultivated and the changes that have taken place in the varietal mix. However, as those measures are calculated from data on named varieties, they do not explicitly take into account the largely unobservable genetic relationships among the cultivated varieties. As in Chapter 6, the levels of crop genetic diversity are calculated using data on variety pedigrees and measure the degree of ancestral commonality, or coefficient of parentage (COP), between a given pair of varieties (Cox et al. 1985; Souza et al. 1994). In this chapter, the coefficient of diversity (COD) is used, where COD = 1-COP. Weighted measures of both the COP (WCOP) and the COD (WCOD) can be derived using the area shares sown to each variety, with the relationship between the two defined as WCOD = 1-WCOP. The WCOD thus measures the diversity of the mix of varieties, as reflected by materials utilized in the breeding process, cultivated in a region in a particular year. Changes over time can indicate the extent to which the diversity in farmers' fields is being eroded or enhanced over time, given the turnover in the mix of cultivated varieties.

In an earlier study, Brennan and Fox (1998) showed that diversity at a national level (at five-year intervals) during 1973-93 remained high, but that the trends varied markedly between states. While a significant improvement in diversity was observed in South Australia and Western Australia, there was a narrowing of diversity in the base of varieties grown in the eastern states over a similar period. Using updated annual data for the period since 1965, we observe little change in overall diversity as measured by the WCOD (Figure 7.7), although a decline is evident in the mid-1960s through early 1970s and again since the mid-1990s. To place these numbers in perspective, Smale (1996) found WCOD for developing countries in 1997 ranged from 44% to 97%, with a mean WCOD of 79% to 83%, very similar to that found here for Australia. The average value of WCOD aggregated across seven major wheat

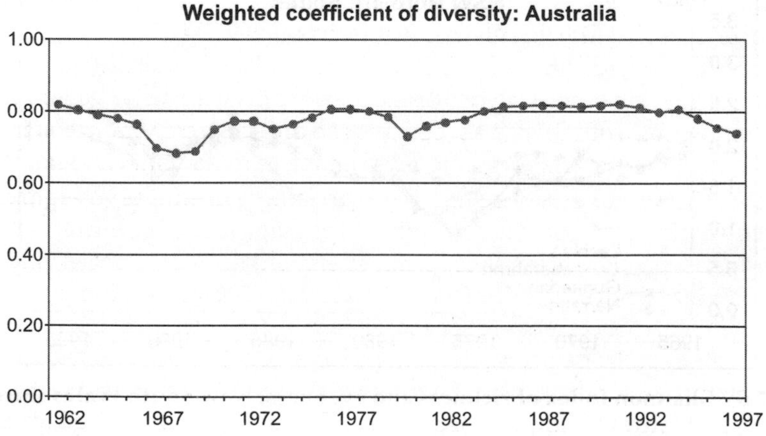

Figure 7.7 Weighted coefficient of varietal diversity: national level.

producing provinces in China from 1982-1997 is relatively high at over 90% (see Figure 6.7). However, the WCOD for Canada, with its stricter emphasis on quality standards for varietal release, was considerably lower than that of Australia. In terms of the world market, Canada as a major wheat exporter bears some similarities to Australia. A plausible explanation for the low level of the Canadian WCOD could be the even stricter quality parameters dictated by their market niche demands than by those in the Australian wheat market.

The increase in varietal richness and evenness indicated by the Margalef and Shannon indices, particularly between 1980 and the early 1990s, is however not reflected to the same extent in an increase in the genetic diversity of the varieties grown. This can be attributed partly to an increasing number of varieties being grown that are closely related in parentage. An increase in varieties released with minor, perhaps cosmetic, differences from other popular varieties has been highlighted in Chapter 4 as one of the likely consequences of the increased commercialisation of breeding programs in recent years, and evidence to support that hypothesis is present here.

Differences, however, can be observed at the state level. Since 1965 (Figure 7.8), there has been a significant decline in WCOD values calculated for both New South Wales and Queensland. The WCOD for New South Wales declined from a value of over 0.80 at the start of the period to 0.50 by 1997, with a particularly noticeable drop in the 1990s. The change in Queensland was less marked until the late 1980s, with the exception of a low period between 1980 and 1984. The decline in the 1990s, however, has been strong, albeit slightly erratic.

Diversity levels as reflected by WCOD have followed different patterns in the other states. Values calculated for Victoria, South Australia, and Western Australia for the end of the period were very similar to those calculated for the 1960s, although some different patterns were observed over that period of time. Initially, Western Australia faced a decline in WCOD as it relied heavily on a single variety, Gamenya, in the late 1960s and 1970s[1]. All three states showed an increased WCOD with the initial uptake of semi-dwarf wheats, in the 1970s in Victoria and South Australia and in the early 1980s in Western Australia. However, while South Australia and Western Australia maintained those levels in the 1990s, Victoria's WCOD fell to around 0.40 in the 1990s. The diversity base for Victoria's varieties was generally lower than that for varieties in the other southern states, reflecting its less varied production environments and the reliance on varieties closely related to the early semi-dwarf variety Condor. The more diverse production environments in South Australia and Western Australia, in contrast, encouraged use of a wider genetic basis in the varieties targeted for those areas. In summary, there is clear evidence of a decline in diversity as measured by coefficients of diversity since the 1970s in New South Wales, Queensland, and Victoria, whereas diversity in South Australia and particularly Western Australia has increased.

[1]*Throughout the 1970s, the variety Gamenya represented 50% or more of the area sown to wheat each year in Western Australia.*

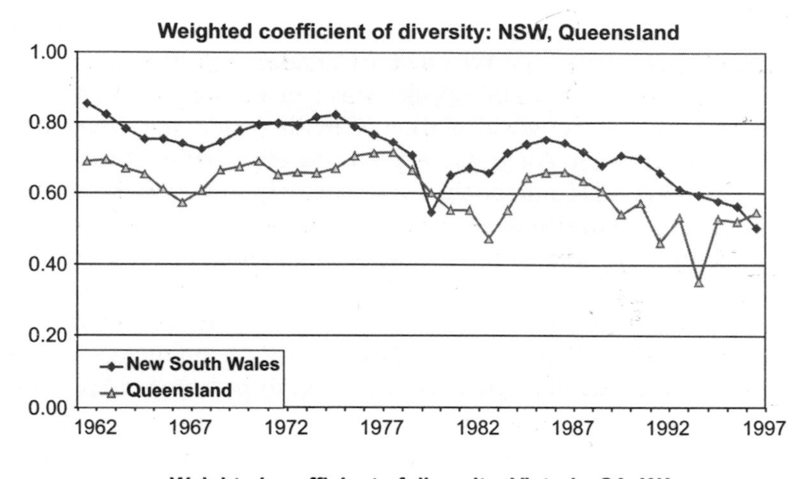

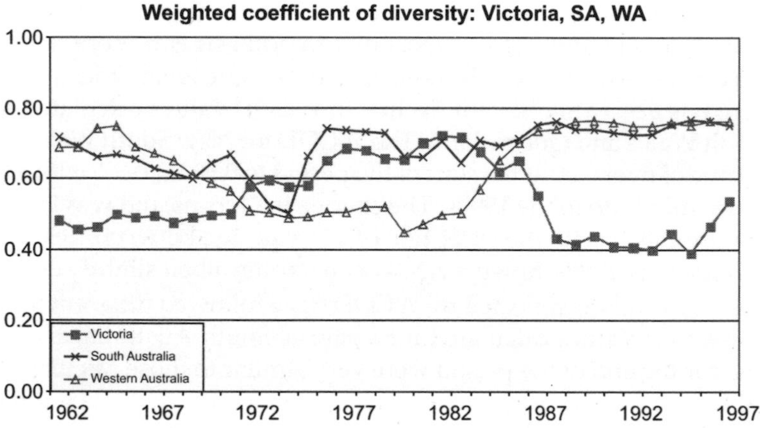

Figure 7.8 Weighted coefficient of varietal diversity: state level.

The increase at the state level in the spatial diversity measures of richness and evenness based on named varieties observed in Figures 7.2 and 7.4 has in fact concealed a decline in diversity as measured with parentage data in the three eastern states. This implies that, despite the move toward an increased number of varieties grown in those states, they have generally been closely related genetically. That has not been the case in South Australia and Western Australia, where the diversity of the parental bases has increased simultaneously with the spatial diversity reflected by their cultivated set of named varieties.

Changes in Latent Diversity at Shire Level

At the shire level, the weighted coefficients of diversity reveal a marked decline in the diversity of the varieties that farmers have been growing since the mid-1970s, particularly in the southern shires (Figure 7.9). Until 1975, WCOD values calculated for the southern shires generally were above 0.6, but

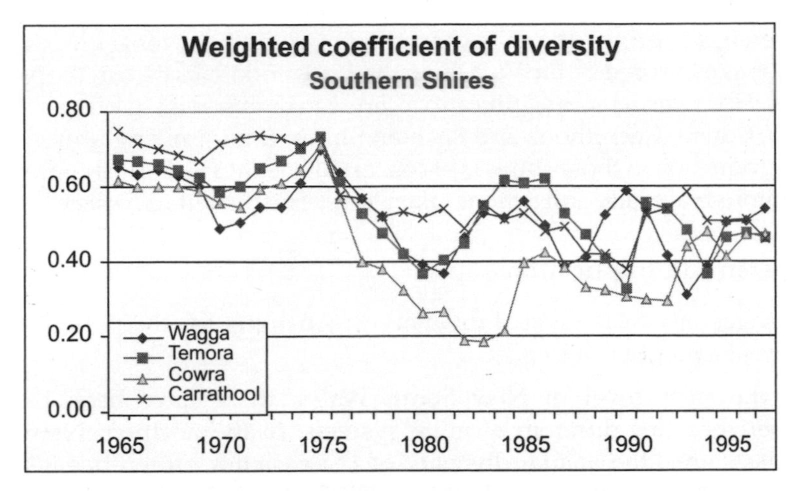

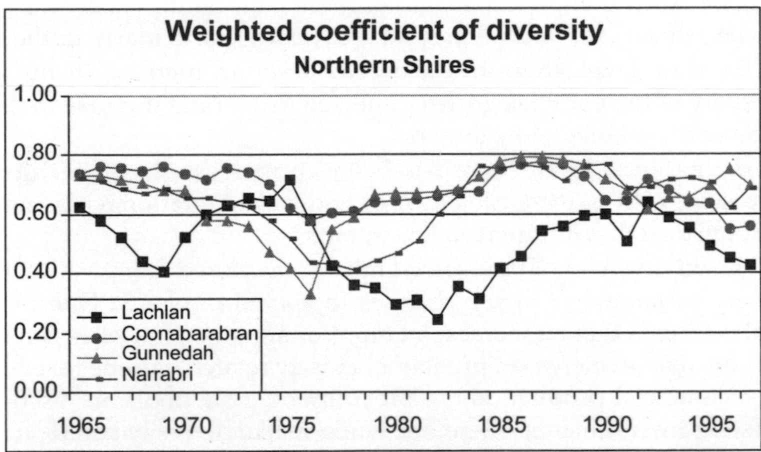

Figure 7.9 Weighted coefficient of varietal diversity, by New South Wales Shire.

subsequently, all four shires dropped below that level in most years. It is likely that the widespread adoption of the first semi-dwarf wheat varieties (Condor and Egret) from the mid-1970s was responsible for the initial fall in the index. Since that time, all southern shires have maintained a WCOD value of around 0.5. Cowra focused heavily on a limited range of varieties, especially in the early 1980s. Although Cowra Shire's mix of varieties as measured by the Shannon index of named varieties appeared diverse in the 1980s, the varieties were in fact closely related and resulted in low levels of diversity as measured by the WCOD. In the north, all shires apart from Lachlan had WCOD values above 0.6 for most of the period. The only time in which they fell below that level was in the 1970s during the initial adoption period of semi-dwarf varieties. In the shire of Lachlan, WCOD levels were similar to those in the southern shires, with a decline in the 1990s reflecting the widespread adoption of two closely related varieties, Janz and Cunningham.

Overall, as a group, the shires have had marginally lower average rates of diversity, as measured by the WCOD, in the 1980s and 1990s than in the previous decades. However, a more significant decline can be observed in Wagga Wagga, Temora, Cowra, Carrathool, and Lachlan Shires. If minimizing uniformity in wheat production in those shires is of concern in the future, these trends need to be monitored carefully and actions taken to reverse them if necessary.

Discussion and Implications

The changes in wheat varietal diversity in Australia for the period 1965 to 1997 reveal a number of trends:

- At the shire level in New South Wales, there have been different experiences for different farming systems. In the northern New South Wales shires, the spatial diversity of the varieties grown has generally increased over that period. In the southern New South Wales shires, there has been a general decline in spatial diversity, particularly in the 1990s.
- At the state level, there has generally been an increase in the spatial diversity of the varieties grown, reflected in a strong increase in the total number of varieties being grown.
- At the national level, there has been an increase in spatial diversity, consistent with an increasing level of regional adaptation replacing broad adaptation across different environments.
- The latent diversity of the varietal mix, as measured using WCOD, is not closely correlated with the changes in spatial diversity. One plausible explanation is that an increase in number of named varieties can simply indicate that an increased number of closely related varieties are released for commercial production or that common gene pools are used across different environments. Therefore, while spatial diversity based on named varieties has increased overall, there was a decline in diversity based on genealogical background in the mix of varieties grown in three of the five states and in six of the eight shires analysed. Brennan et al. (1999) report that commercial and funding pressures on breeders can lead them to increase the number of varieties released and to give less attention to the genetic diversity of and among those varieties (see also Chapter 5). Those trends are supported by these findings.

In the period since 1997, the institutional environment for wheat breeding and diversity in Australia has changed markedly. Over the past ten years, the former public-sector wheat breeding programs have become increasingly commercially-oriented, and many have now been privatised. The seed supply system has also been privatised, intellectual property rights have been widely used to protect all new varietal releases, and research and development programs have been re-structured. These changes are too recent to be revealed in the data in this analysis, but the diversity of Australian wheat may well show that changes occurring since about 2001 have had a significant effect.

At the least, we recognize that the prevailing conditions have changed and that established trends are likely to be altered.

The differences in diversity at different levels of measurement are illustrated in Table 7.1, where each of the diversity measures is shown at the national, state, and shire levels. The measures are higher at the national level than at the state level, and again higher for New South Wales than the average of the shires. The ratio of national to state levels is 2:1 for the Margalef index, 1.4:1 for the Shannon index and 1.2:1 for the WCOD. Ratios of similar magnitude exist between the New South Wales level and the average of its shires. It is clear from the Australian analysis that whatever measure is used, lower levels of measured diversity result when smaller regions are analysed, due to the higher likelihood of similarities among the environments and existing farming systems[2]. As larger regions are analysed, they are likely to include a wider range of production environments and farming systems, and hence varieties, so that diversity measures will be relatively higher, regardless of the data underlying the measure. These relationships are important to bear in mind when making comparisons between different countries or production regions.

Table 7.1 Relationship between diversity measures at different levels.

	Mean index levels, 1965-97		
	Margalef	*Shannon*	*WCOD*
Australia	**5.33**	**2.87**	**0.79**
<u>**States**</u>			
New South Wales	3.16	2.39	0.72
Queensland	2.20	1.81	0.61
Victoria	2.37	1.84	0.56
South Australia	3.10	2.09	0.71
Western Australia	2.66	1.99	0.67
- State mean	***2.70***	***2.02***	***0.65***
<u>**New South Wales Shires**</u>			
Wagga Wagga	1.53	1.46	0.52
Temora	1.77	1.55	0.54
Cowra	1.80	1.40	0.44
Carrathool	1.90	1.72	0.58
Lachlan	1.94	1.60	0.50
Coonabarabran	2.46	1.89	0.69
Gunnedah	2.17	1.81	0.66
Narrabri	1.87	1.82	0.64
- Shire mean	***1.93***	***1.65***	***0.57***
Ratio of Australia: States	1.98	1.42	1.21
Ratio of New South Wales: Shires	1.64	1.44	1.26

[2]*Note that the experience has been different in China (Chapter 6).*

It is apparent from the different levels of spatial diversity analysis that aggregate analysis at the national or state level can obscure the extent to which changes are taking place at a more disaggregated level. Therefore, findings that indicate a general increase in spatial diversity at the aggregate or national level do not necessarily imply that serious issues relating to a loss of varietal diversity do not exist at the local or regional level. The risks of a loss of diversity also need to be identified at the regional or local level in order for spatial and genetic diversity to be managed appropriately, and policies to accommodate these local impacts.

Given that the average wheat area in each shire was approximately 60,000 hectares, it may be difficult to develop cost-effective policies on diversity aimed at the shire level. While some efforts, particularly based on extension advice, are likely to be worthwhile at the shire level, the most appropriate policy level to address diversity issues is likely to be at a higher level of aggregation based on agro-ecological regions that are used as target regions by breeding programs. The size of the appropriate levels will depend on the resources available and the availability of cost-effective policy options to address diversity.

REFERENCES

Brennan, J.P. and A. Bialowas. 2001. *Changes in Characteristics of NSW Wheat Varieties, 1965-1997*. Economic Research Report 8, NSW Agriculture, Wagga Wagga.

Brennan, J.P. and P.N. Fox. 1998. Impact of CIMMYT varieties on the genetic diversity of wheat in Australia, 1973-1993. *Australian Journal of Agricultural Research* 49: 175-178.

Brennan, J.P., D. Godden, M. Smale, and E. Meng. 1999. Breeder demand and utilisation of wheat genetic resources in Australia. *Plant Varieties and Seeds* 12: 113-127.

Cox, T.S., Y.T. Kiang, M.B. Gorman, and D.M. Rodgers. 1985. Relationship between coefficient of parentage and genetic similarity indices in the soybean. *Crop Science* 25: 529-532.

Smale, M. 1996. Understanding Global Trends in the Use of Wheat Diversity and International Flows of Wheat Genetic Resources. Economics Working Paper 96-02. Mexico, D.F.: CIMMYT.

Souza, E., P. Fox, D. Byerlee, and B. Skovmand. 1994. Spring wheat diversity in irrigated areas of two developing countries. *Crop Science* 34: 774-783.

Chapter **8**

Explaining Spatial Diversity in Wheat in Australia and China

M. Smale, E. Meng, J.P. Brennan, and R. Hu[1]

ABSTRACT

The factors that cause spatial diversity to vary across time and environments are examined empirically for both Australia and China. Three measures of diversity (richness, dominance, and evenness) are modeled using panel data for eight shires in New South Wales, Australia (1983 to 1997) and seven provinces in China (1982 to 1995). Using Zellner's seemingly unrelated regression (SUR), a number of factors are identified as significant determinants of spatial diversity across these different states and provinces. Variety characteristics (yield, quality, maturity, height) features of the research system (rate of variety release, level of local adaptation of varieties), and the characteristics of the physical environment (such as moisture regime and irrigation, erosion, salinity, soil variability, multiple cropping) are shown to be important factors in explaining the observed varietal diversity. This improved knowledge of the key factors involved provides a basis for managing diversity in the future.

Before developing policies that influence the spatial distributions of crop varieties or the genes that they embody, we need first to identify the factors that cause them to vary. The spatial distribution of varieties is determined at an aggregate level by the interaction of choices made by individual farmers. Farmers choose varieties based on observable traits and expected performance, rather than on genetic composition that they cannot see. In general, those who grow modern varieties are more reliant on the external supply of genotypes and traits provided through plant breeding programs than those who cultivate landraces, but differences in degree of reliance are also evident within modern systems. Within the range of modern systems, on-farm seed

[1]*This chapter draws heavily on Smale et al. (2003).*

replacement rates vary greatly, and farmers who replace varieties more frequently are more reliant than those saving their own seed. As the economy develops and agriculture is commercialized, farmer demand for specific varieties shifts in focus from their own needs as consumers toward a demand derived from the requirements of industrialized grain processors, export, markets and the preferences of more distant, urban consumers. However, in all crop production systems, including commercialized ones, agro-ecological features of the production environment condition farmers' decisions by affecting the performance of varieties differently. The variety choices of farmers are also constrained by government policies that affect the research and development of varieties, seed sales, and distribution.

In this chapter, we hypothesize that the variation in these indices is affected by the same factors determining farmer's choices of variety. Utilizing time series of three indices of spatial diversity (richness, dominance, and evenness) constructed from panel data on the area shares sown to wheat varieties in New South Wales (NSW), Australia (1983–97), and seven provinces in China (1982–95), we test our hypotheses econometrically using Zellner's seemingly unrelated regression (SUR).

The two regions of study provide some essential points of comparison. The wheat varieties grown in China are produced for both commercial and subsistence purposes. Approximately 19.2 million hectares of bread wheat were grown in 1997 in the seven provinces included in this study (Anhui, Hebei, Shanxi, Jiangsu, Shandong, Henan and Sichuan). Most of the wheat grown in these provinces has facultative[2] and winter growth habit, although wheat with spring habit is cultivated in parts of Anhui and Jiangsu provinces and most of Sichuan province, where they are planted in the autumn. Wheat area by habit is difficult to estimate in China since both spring and winter habit varieties may be planted in the fall.

The wheat varieties grown on about 2.8 million hectares in New South Wales are generally bread wheats with spring growth habit that are grown exclusively for commercial purposes. In the early 1990s, NSW produced over 25% of the total wheat crop of Australia on about 22% of national wheat area (Brennan 1999). The average area of wheat grown on specialist cropping farms was over 400 hectares and on mixed livestock and cropping farms slightly under 150 hectares. Specialist cropping farms produce a number of other crops on large areas in addition to wheat, as well as some livestock (see Chapter 4).

We discuss the conceptual basis for the econometric estimation used to explain variation in spatial diversity in the next section. The spatial diversity indices calculated for Australia and China are briefly described in the

[2] *Facultative wheats are intermediate in vernalization requirements to spring and winter wheats. They are often planted in autumn, like winter wheats, in areas where winter temperatures are relatively warm.*

subsequent section, along with a description of data sources and variables. Regression results are then presented and interpreted, followed by some conclusions.

DETERMINANTS OF SPATIAL DIVERSITY

Factors Affecting Spatial Diversity

The level of diversity in the wheat crop growing in farmers' fields is the outcome of allocation decisions among varieties in their cultivated area each season. The quantity of seed the farmer sows for a given variety, which can also be expressed as a proportion of that farm's wheat area, represents the farmer's constrained demand for that variety. Factors that determine wheat production decisions of individual farmers are also the factors that determine variety shares in aggregated regions (Brennan et al. 1999).

Farmers' choices reflect their demand for the variety traits that confer economic value (Barkley and Porter 1996). In a commercial system like Australia's, such traits include days to maturity, expected yield, grain quality, and resistance to lodging (Brennan 1988). These traits are fixed from the viewpoint of the farmer, though they are malleable to change over time through plant breeding. Seed-to-grain price ratios often do not matter for decisions among varieties, since prices and costs are often the same for the same class of wheat. Yield differences and price premiums for grain quality determine relative profitability.

Household food self-sufficiency is an important means of minimizing food expenditures in China for rural households at all income levels. Survey data show that over 80% of basic food grains consumed on farm were produced by the household (Gale et al. 2005). In this type of semi-commercial system, yield differences are important in determining the relative value of varieties to households. However, relative value also depends on whether the households are net sellers or net purchasers of wheat as well as household characteristics that affect their on-farm wheat consumption and access to markets. Since many of these factors are household specific, they are more difficult to measure at an aggregated regional or provincial level.

Farmers' choices are constrained by the supply of seed of their preferred varieties. With a self-pollinating crop such as wheat, new germplasm is supplied to farmers, particularly in commercial systems, as the product of public and private breeding programs. The supply of varieties is determined by a complex mix of factors, including but not limited to past investments in research, the flow of germplasm and varieties from other programs, and policies affecting breeding objectives, variety release, and seed sales and distribution.

Agro-ecological features of the crop production zone such as soils and rainfall condition the choices of farmers in the region. Though systems of modern wheat varieties do not respond to the selection pressures of the environment as would systems composed of landraces, the heterogeneity of the production environment influences the performance of the genetic materials that the seed system provides. We might hypothesize, for example, that difficult growing environments might lead farmers to choose a broader set of varieties to suit various soil and seasonal niches. In the aggregate, a more heterogeneous and variable environment might display a greater mix of varieties in which none is able to dominate, unless that variety were widely adapted. Features of the production environment are not affected in any significant way by the specific conditions of any one farm or by the deliberate actions of any one farmer.

Spatial diversity indices for richness (Margalef), inverse dominance (Berger-Parker), and evenness (Shannon) based on named wheat varieties grown by farmers in NSW and China have been discussed in detail in Chapters 2, 6 and 7. The indices are constructed from the proportional distributions of area by named variety, or area shares. We hypothesize that the economic determinants of farmers' variety choices explain the variation in these indices. Agro-ecological factors of the crop production zone, the supply of variety-specific traits that can be observed by farmers, and the policies that affect the distribution of new wheat varieties to farmers are exogenous factors explaining spatial diversity among modern wheats in Australia and China.

Specification of Estimating Equations

Since the richness, inverse dominance and evenness indices express different spatial diversity concepts, each was specified separately as a function of a set of related but distinct variables that determine the demand for and supply of varieties. This specification reflects the hypothesis that determinants of spatial diversity operate differently depending on the diversity concept. In the most general form, the three equations in the systems can be represented as:

$$D^r = D^r\,(\mathbf{X}^r \mid \mathbf{S}, \theta, \mathbf{Z}) \tag{8.1}$$

$$D^d = D^d\,(\mathbf{X}^d \mid \mathbf{S}, \theta, \mathbf{Z}) \tag{8.2}$$

$$D^e = D^e\,(\mathbf{X}^r, \mathbf{X}^d \mid \mathbf{S}, \theta, \mathbf{Z}) \tag{8.3}$$

The richness (D^r), inverse dominance (D^d), and evenness (D^e) of the wheat varieties grown by farmers in a region over time is determined by the observable characteristics of the varieties that have economic value to them (vector **X**), factors that affect the supply of varieties and germplasm (vector **S**), parameters of the diffusion curve (θ), and agro-ecological factors (**Z**).

The variables used in the regression models for NSW and China are defined in Table 8.1. The dependent variables in both models are Margalef, Berger-Parker, and Shannon indices of spatial diversity constructed from data on variety area shares by province or region and year.

Table 8.1 Definitions of variables used in regressions

Variable	*Definition*
Australia (shire and year)	
D^r	Margalef richness index for wheat varieties grown
D^d	Berger-Parker dominance index for wheat varieties grown
D^e	Shannon evenness index for wheat varieties grown
$\mathbf{X}^r$	Average relative yield potential of wheat varieties grown
	Average bread-making score of wheat varieties
$\mathbf{X}^d$	Relative yield potential of variety with highest area share
	Maturity class of variety with highest area share
	Height class of variety with highest area share
	Bread-making quality of variety with highest area share
S	Number of varieties released in past five years
	Recommended varieties as proportion of varieties grown
	Varieties bred locally as proportion of varieties grown
	1=regulated market period to 1989, 0 otherwise
$\mathbf{Z}$	Shannon evenness index of soil types relevant to wheat production
	Average rainfall in mm from April to October
	Probability of being able to sow early = 1 if rainfall from April 10–30 > 30 mm, 0 otherwise
	Probability of having to sow late = 1 if rainfall from April 10–30 < 30 mm and rainfall in June > 15 mm, 0 otherwise
	0–1 variables for Carrathool, Coonabarabran, Cowra, Lachlan, Narrabri, Temora and Gunnedah
θ	Lagged area-weighted average age of varieties
China (province and year)	
D^r	Margalef richness index for wheat varieties grown
D^d	Berger-Parker dominance index for wheat varieties grown
D^e	Shannon evenness index for wheat varieties grown
$\mathbf{X}^r$	Average yield potential of wheat varieties
	Range in days to maturity among wheat varieties
	Range in height among wheat varieties
$\mathbf{X}^d$	Expected yield of variety with highest area share
	Days to maturity of variety with highest area share
	Height of variety with highest area share
	Protein content of variety with highest area share
S	Crop research expenditures, in million yuan (1985=1), lagged by four years
θ	Lagged area-weighted average age of varieties
$\mathbf{Z}$	0–1 variables for Anhui, Hebei, Henan, Jiangsu, Shandong, Shanxi saline area
	Area affected by drought
	Area affected by flood

Eroded area
Multiple cropping index
Ratio of irrigated area to crop area
Interaction term of ratio of irrigated to crop area with maize-wheat region
1=1982-1984, household reform system, 0 otherwise
1=1991-1995, market liberalization, 0 otherwise

The variety traits (**X**) hypothesized to be associated with spatial diversity of the varieties sown by shire or province and year are relative yield potential, maturity, height and grain quality. Variety traits may be expressed or measured differently in each of the three equations as a reflection of the different spatial diversity concept represented in each equation and of quantitative or qualitative nature of the variety-specific data used to construct the variables. For example, in NSW the height and maturity of varieties is recorded as a class, while data on relative yield potential and bread-making quality are quantitative. Since there is no meaningful way of summarizing height and maturity classes across varieties, these data are included as explanatory variables only in the dominance equation, as the class of the dominant variety. In China, maturity and height variables are quantitative, and their ranges are included as explanatory variables in the richness equation. Protein content, which we hypothesize to be related to grain quality for farmers who either consume or sell, was not available for most varieties, and is included only for the leading variety in the inverse dominance equation. As evenness among varieties by definition consists of elements of both richness and relative abundance, both sets of the variety-specific factors ($\mathbf{X}^r$, $\mathbf{X}^d$) are used in the evenness estimation.

The supply of wheat varieties (**S**) is measured in NSW by the total number of varieties released in the preceding five years and by the proportion that were bred locally. Other supply-related variables include the proportion of varieties grown that are recommended or approved by NSW Agriculture, and an indicator variable to capture the change in policy regime from the regulated period (pre-1990) to the deregulation period that began in 1990. For China, policy indicator variables are used to mark changes in policy regimes with both the household reforms of the early reform period beginning in 1978 and a period of increased market liberalization beginning in the early 1990s (Chapter 3). A variable for the overall level of government expenditures in crop research is used to represent the supply of varieties (**S**).

The vector **Z** includes regional indicator variables for shires or provinces, as well as variables representing features of the agro-ecology and farming system. In NSW, we used an index of evenness in the distribution of soil types, constructed from geographically referenced data as the Shannon index over major soils classes relevant for wheat production (Brennan and Bialowas 2001). Moisture regimes are measured with three variables constructed from rainfall data: (1) average growing-season rainfall from April to October; (2) the possibility of being able to sow early; and (3) the possibility of having to sow late. In China, agro-ecological variables include individual estimates of the extent of land area affected by droughts, floods, erosion, and salinity. Variables reflecting the coverage of irrigation systems and cropping intensity are also included in the Chinese equations.

In both Australia and China, on-farm seed supplies and a lagged response to variety release influence past area allocation decisions. There is some inertia in changing varieties because most farmers save their seed from year to year and purchased seed accounts for only a small percentage of sown wheat area in any given year. Diffusion information, including the initial adoption lag and the length of the adoption period, incorporates all these influences, and is summarized in the area-weighted average age of varieties (Brennan and Byerlee 1991). This variable is lagged by one year to ensure exogeneity.

Data Sources

Data sources for both China and Australia are numerous. In NSW, eight shires were selected to represent farming systems across the state: Wagga Wagga, Temora, Cowra, Carrathool, Lachlan, Coonabarabran, Gunnedah, and Narrabri (Brennan and Bialowas 2001). All varieties grown in each shire in the period 1983–97 were identified. The result was a set of 59 varieties covering at least 80% of the area sown to known varieties in each shire in each year from 1983 to 1997[3]. From data on the year of release of each variety and the breeding program developing the variety, the varieties that were bred "locally" for each shire were identified. The local varieties for southern shires (Wagga, Temora, Cowra, and Carrathool) were those released by the Wagga Wagga and Temora breeding programs, while for the other shires "local" varieties were those released by the breeding programs at Narrabri, Tamworth or Toowoomba. From 1983 to 1988, variety share data were the percentage of the area of wheat sown to each variety in each shire. From 1990 to 1997, the only available variety data were the percentage of wheat receivals at local silos by the Australian Wheat Board (AWB). To provide comparable data, a representative silo was selected in each shire, and the variety share data for that silo taken to represent that of the shire as a whole. Data for 1989 were interpolated from 1988 and 1990 data.

Variety yield data were taken from all advanced trials conducted by NSW Agriculture from 1982 to 1998. All sites in the selected shires in those years were combined, and the trial yields analysed to provide a ranking of all varieties against standard varieties. NSW Agriculture's annual list of recommended varieties, based on the latest yield, quality, disease, and marketing information, was also used to classify varieties into three maturity types: late maturing, mid-maturing, and early maturing. Bread-making quality

[3]*The detailed sources of these data are provided in Brennan and Bialowas (2001).*

is reflected by a score assessing varieties on a scale of 1 to 10, with 10 being perfect for bread making. Data on the morphological characteristics of each variety were used to construct additional variables. Finally, the soil types present in the arable portions of each shire were identified using spatial imaging, and the area and percentages of each type was obtained for each shire.

A policy variable was developed to account for effects of Australian wheat-industry policy on varietal-choice decisions. After consideration of the history of policy development in the wheat industry (see Chapter 4), the following time periods were assessed as being important in determining varietal choice in the period of this analysis:

- *1983-89.* A regulated period in which substantial classification of varieties was used by the AWB, which controlled the marketing of all Australian wheat.
- *1990-97.* A deregulated period in which the power of the AWB to control wheat marketing in Australia was restricted to exports, and the domestic market was fully deregulated.

For China, panel data on input and output prices, expenditures, environmental conditions, and agricultural research investment were assembled for the provinces of Anhui, Hebei, Henan, Jiangsu, Shanxi, Shandong, and Sichuan from 1982 to 1995. China's statistical and agricultural yearbooks were the primary sources for data on area sown to wheat and production. Additional information on variety area shares was taken from relevant years of the Ministry of Agriculture (MOA) publication, *Statistical Compendium, China's Major Varieties*, and from interviews with personnel at the Henan and Shandong Provincial Seed Management Stations. Data on agricultural research investment were obtained from the National Science and Technology Bureau. Total agricultural research expenditures were adjusted by the share spent specifically on crops research and lagged by four years, to represent the time spent finishing varieties. Yield and trait data refer to the time of official variety release and were obtained by a search of publications on wheat varieties and breeder surveys conducted by the Center for Chinese Agricultural Policy. Finally, data on environmental variables were drawn from Ministry of Water Resources (MWR) yearbooks.

Econometric Results

The same errors that affect the richness of the varieties planted in any season are also likely to affect their relative abundance and the evenness of their distribution over a geographical area. Zellner's seemingly unrelated regression (SUR) model exploits the underlying relationships in the errors among equations by estimating them jointly. When each dependent variable is a function of the same explanatory variables, the estimation is equivalent to Ordinary Least Squares. When the three dependent variables are functions of the different explanatory variables but are related through their error structures, statistical efficiency gains may be made through SUR estimation. The greater the correlation of the disturbances among equations, or the more distinct the matrices of explanatory variables, the greater the efficiency gains from running the equations jointly (Greene 1997).

Model for Shires in New South Wales

Results of the SUR estimation[4] for NSW are shown in Table 8.2. All three equations are statistically significant at the 1% level, as indicated by the log-likelihood ratio test. Differences are apparent among the regressions in the significance and interpretation of the effects of individual explanatory factors. While economic concepts have been used to motivate the specification of the regression equations, the direction of marginal effects is not predicted *a priori* by theory, therefore all hypothesis tests were two-tailed tests.

No trade-off is apparent between any of the diversity indices and the yield potential of varieties. The richness and the evenness in the spatial distribution of the varieties grown by farmers are positively related to their average relative yield potential. Though we might hypothesize that higher yields are associated with the cultivation of fewer, higher-yielding varieties over a larger proportion of area, this result suggests that no annual losses were associated with greater spatial diversity of modern varieties over the period since 1983 in NSW. Breeding for successive improvements in yield performance in trials and seeking to ensure a rich and even distribution of varieties from year to year do not appear to have been conflicting goals.

Other variety characteristics that are observable to farmers also explain variation in the spatial diversity of wheat varieties grown in NSW. Since the Berger-Parker index is the inverse of the maximum area share of the varieties sown in any given year, a negative sign indicates a positive relationship with a variety's dominance. Thus, the higher the bread-making quality of the dominant variety and the shorter its height, the greater the extent of its dominance. Conversely, the later the maturity, the more equitable the spatial distribution of varieties. Later maturity also contributes to greater evenness because evenness refers to the proportional abundance across all varieties and late maturing varieties do not dominate as a class.

Factors related to the effective supply of varieties are also important determinants of spatial diversity among the modern wheats grown in NSW. The greater the relative proportion of locally bred varieties, the greater the richness and evenness among varieties, and the magnitude of this effect is relatively large. In contrast, a higher proportion of recommended varieties

---

[4]*Regressions were run using LIMDEP 7.0. In the SUR (iterative GLS) regression, the test of significance of individual coefficients is the Z statistic. The significance of each equation was evaluated with a log-likelihood ratio test comparing regression on a constant with the hypothesized regression model.*

Table 8.2 Results of generalized least squares regression (SUR) of wheat diversity indices in New South Wales.

Explanatory variable	*Richness coefficient*	*S.E.*	*Dominance^{-1} coefficient*	*S.E.*	*Evenness coefficient*	*S.E.*
Constant	–0.00440	0.862	0.894	1.77	–0.166	0.840
Average relative yield potential	0.0114+	0.00622			0.00907+	0.00523
Relative yield of dominant variety			0.00222	0.00450	–0.00165	0.00112
Average bread-making quality	–0.0416	0.519			–0.328	0.0442
Bread-making quality of dominant variety			–0.267+	0.163	–0.086*	0.0429
Later maturity of dominant variety			–0.00497	0.145	0.0732*	0.0360
Taller height of dominant variety			0.109+	0.0610	0.0259	0.0174
Proportion locally bred	0.395+	0.206	0.471	0.723	0.528*	0.232
Proportion recommended	–0.403*	0.0651	–0.681*	0.244	–0.399*	0.0763
Varieties releases in last 5 years	0.0202*	0.00981	–0.0106	0.0355	0.0465*	0.0113
Regulated market period	–0.105*	0.0579*	0.696*	0.215	0.132*	0.0673
Rainfall	–0.000491*	0.000233	–0.000746	0.000807	–0.000337	0.000270
Soil type evenness	–0.00994	0.299	1.72+	1.02	0.868*	0.335
Sow early	0.00735	0.0403	–0.264+	0.146	0.00148	0.0462
Sow late	–0.102+	0.0649	0.176	0.239	0.0129	0.0750
Lagged area-weighted average age	0.00725	0.0194	0.113+	0.0674	0.0369+	0.0220
Carrathool	0.0929	0.0853	–0.115	0.313	–0.000892	0.0987
Coonabarabran	0.452*	0.112	–0.977	0.342	0.622*	0.121
Cowra	–0.119	0.146	0.538	0.455	0.109	0.160
Gunnedah	0.381*	0.158	1.89*	0.423	0.846*	0.164
Lachlan	0.453*	0.0875	0.823*	0.331	0.402*	0.103
Narrabri	0.191	0.133	1.91*	0.345	0.494*	0.138
Temora	0.214	0.161	1.19*	0.564	0.578*	0.182
Value of log-likelihood ratio	86		57		99	
Number of observations	104		104		104	

* Statistically significant at 5% level of significance with two-tailed Z statistic.

+ Statistically significant at 10% level with two-tailed Z statistic.

among those grown by farmers is associated with greater dominance in the leading variety and a less even set of variety area shares. Adoption lags and slower diffusion patterns, as reflected in the lagged area-weighted average age of varieties, reduce the dominance of any single variety and improve spatial evenness among varieties. These results are intuitive when we consider that, as older varieties or varieties that are no longer recommended shift gradually out of production, the minor areas they occupy serve to enhance diversity from a spatial perspective. A higher rate of variety release also enhances richness and offsets the negative effects of recommended varieties on the equity of their spatial distribution. Prior to deregulation in the wheat market in 1990, the richness of the varietal mix was lower and the inverse dominance and evenness higher than they have been since 1990. These findings indicate a significant shift from a scenario with lower per unit diversity but less concentration in dominant varieties and relatively more even distribution to a scenario with a larger pool of material drawn upon to develop individual varieties but more concentration in fewer dominant varieties.

Physical features of the production environment are also important in explaining variation in the spatial diversity of the wheat varieties grown in NSW. A higher average level of precipitation, a characteristic of a relatively more optimal growing environment, all else equal, is negatively associated with richness of wheat varieties, as is the possibility of having to sow late. A better moisture regime may mean that more farmers choose to grow fewer varieties, while a delay in the sowing time implies that fewer varieties are suitable to the more unique requirements of a short growing period. The evenness in the distribution of soil types relevant to wheat production reduces the dominance of the leading variety and enhances the evenness of variety area shares. This finding is consistent with the notion that there is some soil-specificity in the performance of varieties. Wheats grown in the shires Gunnedah, Lachlan, Narrabri, Temora, and Coonabarabran are in one respect or another more spatially diverse than those grown in the shires of Carrathool and Wagga Wagga.

Model for China Provinces

Results of the SUR regression for seven major provinces in China over the period 1982 to 1997 are shown in Table 8.3. Log-likelihood ratio tests confirm that each individual regression is significant at the 1% level. As in NSW, variety characteristics other than yield potential are significantly associated with the richness, dominance, and evenness in the spatial distribution of wheat varieties grown. Later maturity, in particular, is associated with a lower area share of the dominant variety. The greater their range in maturity period and height, the greater the richness and evenness of the wheat varieties sown by farmers, implying that the availability of alternatives in variety traits is both useful and utilized.

Table 8.3 Results of generalized least squares regression (SUR) of wheat diversity indices in seven provinces of China.

Explanatory variable	*Richness coefficient*	*S.E.*	*Dominance^{-1} coefficient*	*S.E.*	*Evenness coefficient*	*S.E.*
Constant	−21.943	7.37	−73.77*	28.89	−15.08	4.56
Average yield potential	−0.00434*	0.00246			−0.00231	0.00152
Yield potential of dominant variety			0.00229	0.00223	0.000272	0.000300
Range in maturity	0.00284*	0.00994			0.000575	0.000582
Range in height	0.0205*	0.00537			0.0128*	0.00311
Later maturity of dominant variety			0.0383*	0.0214	0.0000650	0.00292
Protein content of dominant variety			−0.212	0.148	−0.0205	0.0192
Taller height of dominant variety			0.0163	0.0248	0.00160	0.00321
Salinity	0.00309	0.00319	−0.0460*	0.0120	−0.00252	0.00191
Erosion	0.000605*	−0.000291	0.00209*	0.00115	0.000475*	0.000181
Flood	−0.0000151	0.0000647	0.000143	0.000239	0.0000245	0.000037
Drought	0.0000369	0.0000580	0.000148	0.000222	0.0000273	0.0000347
Multiple cropping index	3.63*	0.806	10.97*	3.51	2.47*	0.528
Ratio of irrigated to total cultivated area	2.22+	1.26	−19.80*	5.00	−0.615	0.797
Interaction of irrigation ratio and maize-wheat region	1.18	1.82	17.1*	6.99	0.973	1.09
Lagged crop research expenditures	−0.000385	0.0278	−0.000323	0.00105	0.0000222	0.000167
Lagged area-weighted average age	0.0985*	0.0355	0.876*	0.134	0.113*	0.0211
Household reform system	0.0561	0.177	−1.59*	0.645	−0.259*	0.106
Market liberalization	0.261*	0.119	−0.165	0.440	−0.066	0.0718
Anhui	14.18*	6.82	56.64*	27.17	11.84*	4.25
Hebei	10.09	7.768	85.39*	30.61	13.00*	4.81
Henan	10.21	7.35	70.50*	29.00	11.83*	4.55
Jiangsu	12.37	8.86	89.54*	34.99	14.23*	5.49
Shandong	10.87	8.29	84.18*	32.76	13.58*	5.13
Shanxi	11.25*	5.12	45.64*	20.847	10.13*	3.24
Value of log-likelihood ratio	103		79		105	
Number of observations	91		91		91	

* Statistically significant at 5% level of significance with two-tailed Z statistic.
+ Statistically significant at 10% level with two-tailed Z statistic.

In contrast to NSW, average yield potential of the varieties sown by farmers is negatively associated with richness. In other words, when fewer varieties were grown per unit of area, these varieties had higher yield potential. This result could reflect a history of government control of the research system and its heavy prioritization on yield for food security reasons. Although in the period covered by the data, government policies and their possible effects on farm behavior have been complex, the household reform policies instituted during the late 1970s allowing farmers the flexibility to sell surplus output after the fulfilment of their official production quotas and the increasing liberalization of markets in the 1990s (Chapter 3) have likely created incentives for farmers to prioritize production output. Regression results demonstrate that the dominance of the leading variety was greater, and the distribution of varieties less even, during the latter part of the period covered by the data. With the new incentives and income opportunities, farmers have apparently concentrated the area sown on the highest-yielding variety or the variety most suitable for sale. The market liberalization that has occurred since 1990, however, is also thought to have improved the supply of seed relative to the intervening, 1985–90 period. Regression results are consistent with this hypothesis, indicating that richness, or the number of varieties per unit of area, was greater in the more recent period. Expenditures on crop research are not associated statistically with any of the diversity indexes, perhaps because the variable is measured too broadly and is therefore only indirectly related to the supply of varieties. In China, as in NSW, a higher area-weighted age of varieties contributes to greater spatial diversity, since the older varieties occupy minor areas as farmers gradually discard them in favor of newer materials.

Similar to the case of NSW, the regressions for China also illustrate the importance of agro-ecological factors in explaining the spatial diversity among modern wheat varieties. The greater the area affected by salinity, the greater the dominance of the most popular variety, perhaps explained by a comparatively better performance on these soils by a limited number of varieties due to the limited genetic pool for salinity tolerance. However, greater richness, less dominance, and more even distributions of wheat varieties are found where there is more eroded crop area, perhaps due to the absence of a single variety that is better suited for production in a larger variation of fragile growing conditions. The greater the irrigated area as a proportion of all cultivated area, the more the leading variety dominates wheat area. The interaction effect of irrigated area with the maize-wheat region dampens the dominance of the most popular wheat variety. The multiple cropping index reflecting cropping intensity is also associated with a higher level of spatial diversity as represented by richness, inverse abundance and evenness measures. Finally, the econometric results confirm that all provinces are significantly more diverse than Sichuan province, in terms of both the inverse dominance and evenness indices.

Conclusions

To the extent that the spatial distribution of wheat varieties has an effect on crop productivity, it is likely to be associated with a positive economic value. When spatial diversity indices are constructed from variety area shares, we hypothesize that their variation can be explained by economic factors related to the supply and demand for varieties and the physical features of the production environment. Econometric estimation of reduced form equations in a SUR system supports these hypotheses for two contrasting production systems in China (1982 to 1995) and Australia (1983 to 1997). The importance of variety characteristics such as bread-making quality, maturity, and height in explaining the diversity of wheat varieties grown by farmers is evident in both systems. While there is an apparent trade-off between yield potential and the richness of wheat varieties grown in China, higher yield potential is consistent with greater richness and evenness in the spatial distribution of wheat varieties grown in NSW.

The two sets of regressions provide information of general relevance for agricultural research policy. In both systems, a slower rate of variety turnover in the field is positively related to wheat diversity, since older varieties occupy minor shares as farmers gradually replace them with newer varieties. This result suggests a possible yield trade-off over the longer term, since variety turnover is a principal defence mechanism against pathogens that evolve the ability to overcome the disease resistance bred into modern varieties. In NSW, a more rapid rate of variety release and a higher proportion of locally bred material enhance spatial diversity. The deregulation of the Australian wheat market has been associated with a reduction in richness of the variety mix, but an increase in the level of spatial diversity by other measures. In the Chinese data, the effect of market liberalization on the richness of the wheat varieties grown by farmers is positive, while the period of the household reform system is associated with greater dominance of leading varieties and more uneven distribution of variety area shares.

Both sets of regressions confirm the major role of the physical environment in determining spatial diversity. Moisture regimes and the probability of having to sow late influence the number of varieties grown per unit of area in NSW. Evenness in the distribution of soil types is related to the evenness of area allocation among wheat varieties, as well as to the dominance of the leading wheat variety. Erosion, salinity, irrigation, and the multiple cropping index are key factors in the Chinese setting. Shire and province characteristics are important in both NSW and China, respectively.

The results of this analysis provide a basis for an improved understanding of the factors influencing spatial diversity in wheat. We have been able to test some of the hypotheses discussed in Chapters 3 and 4 related to the role of policy and environmental factors, as well as the pivotal role of the supply of diversity through plant breeding programs.

REFERENCES

Barkley, A.P. and L.L. Porter. 1996. The determinants of wheat variety selection in Kansas, 1974 to 1993. *American Journal of Agricultural Economics.* 78: 202-211.

Brennan, J.P. 1988. *An Economic Investigation of Wheat Breeding Programs.* Agricultural Economics Bulletin No. 35, Department of Agricultural Economics and Business Management, University of New England, Adelaide.

Brennan, J.P. 1999. Efficiency in wheat improvement research: A case study of Australia. In *The Global Wheat Improvement System: Prospects for Enhancing Efficiency in the Presence of Spillovers,* M.K. Maredia and D. Byerlee (eds.). CIMMYT Research Report No. 5. Mexico D.F.: CIMMYT.

Brennan, J.P. and A. Bialowas. 2001. *Changes in Characteristics of NSW Wheat Varieties, 1965-1997.* Economic Research Report 8, NSW Agriculture, Wagga Wagga.

Brennan, J.P. and D. Byerlee. 1991. The rate of crop varietal replacement on farms: Measures and empirical results for wheat. *Plant Varieties and Seed* 4: 99-106.

Brennan, J.P., D. Godden, M. Smale, and E. Meng. 1999. Variety choice by Australian wheat growers and implications for genetic diversity. Contributed paper presented at the 43rd annual conference of the Australian Agricultural and Resource Economics Society, Christchurch, New Zealand.

Gale, F., P. Tang, X. Bai, and H. Xu. 2005. *Commercialization of Food Consumption in Rural China.* Economic Research Report Number 8. Washington, D.C.: U.S. Department Agriculture, Economic Research Service.

Greene, W.H. 1997. *Econometric Analysis.* Third edition. Upper Saddle River, New Jersey: Prentice-Hall.

MOA (Ministry of Agriculture). 1986-1997. *Statistical Compendium, China's Major Varieties.* Unpublished Statistical Volume. Beijing, China: Ministry of Agriculture Press.

MOA (Ministry of Agriculture). 1980-2002 (various issues). *Zhongguo Nongye Nianjian [China Agricultural Yearbook].* Beijing, China: Ministry of Agricultural Press.

MWR (Ministry of Water Resources). 1988-2000 (various issues). *Water Conservation Yearbook of China.* Beijing: China's Water and Electric Press.

Smale, M., E. Meng, J.P. Brennan, and R. Hu. 2003. Determinants of spatial diversity in modern wheat: Examples from Australia and China. *Agricultural Economics* 28: 13-26.

Chapter 9

Determinants of Wheat Diversity in Chinese Household Farms

P. Qin, E. Meng, J. Huang, and R. Hu

ABSTRACT

Decisions made by the household ultimately determine the foundations for crop diversity at all levels of aggregation. A survey of households in three wheat-producing provinces (Shandong, Shanxi, and Gansu) in China provides the basis for an analysis of the factors determining diversity at the household level. Three definitions of diversity for named varieties are tested in the analysis. The results show that household characteristics such as farmer age and education do not have a significant effect on diversity, but that the household's ability to bear risk is significant. Household consumption patterns, its degree of commercialization, characteristics related to household land, and the supply of available varieties all play a significant role in determining household diversity. The analysis provides a solid basis for determining policies that will affect diversity at the household level.

Aggregate-level diversity outcomes are closely linked to decisions made at the household level. While constraints affecting farmers' decisions may not necessarily originate at the household level, a full understanding of aggregate diversity and its implications requires consideration of the factors that are most influential at this level. Decisions made by the household ultimately provide the foundation for crop diversity outcomes at all levels of aggregation. In this chapter, using household survey data from three major wheat-producing provinces in China—Shandong, Shanxi, and Gansu—we explore household-level determinants of wheat diversity, thus enabling a better understanding of the crop diversity maintained in farmers' fields and a more empirically-based indication of the household factors influencing crop diversity in China.

This chapter begins with a brief review of the relevant literature and a discussion of the conceptual framework used in the analysis. Next, we present the data and methodology for the econometric estimation of the factors determining household-level diversity. Finally, we discuss estimation results and policy implications.

HOUSEHOLD DECISIONS AND GENETIC DIVERSITY OUTCOMES

Household Decision-making Framework

Major factors underlying technology adoption decisions by households have been examined in detail in the economic literature. Numerous studies identify risk, in particular, as a key factor in household decisions to cultivate multiple crops or multiple varieties of a given crop (Lee 2005; Morduch 1995; Ames and Reid 1993; Just and Zilberman 1983; Feder 1980). Other factors, however, also play an important role in household decisions. The importance of diverse agro-climatic conditions, in particular soil and land quality, and market environment have been emphasized (Bellon and Taylor 1993; Fafchamps 1992; de Janvry et al. 1991). In addition, cultural considerations, such as the use of particular crops and crop varieties for specific dishes and festivals, influence household cultivation choices (Brush 2004; Smale et al. 2001). Cultural factors can be especially important in centers of crop origin or domestication where traditional varieties continue to be cultivated.

The linkages between household level decisions and crop diversity have increasingly become the focus of research, particularly in centers of diversity and domestication (Smale 2006; Van Dusen and Taylor 2005; Smale et al. 2001; Meng et al. 1998; Brush et al. 1992). Using household survey data and household-level diversity indices calculated from morphological characteristics of cultivated wheat varieties, Meng et al. (1998) explore household incentives for the continued cultivation of traditional wheat varieties in Turkey and implications for the feasibility of on-farm conservation of wheat genetic resources. In their theoretical model of household variety choice, the utility derived from cultivating wheat varieties is maximized taking into consideration factors affecting the household's ability to bear risk, plot-level production environment, and market access. Incorporating qualitative information from Turkey, the model assumes that diversity levels maintained by the households are largely determined once the household decides upon its optimal mix of varieties. An explicit decision regarding the level of crop diversity in the household thus does not exist as a separate decision from the variety choice decision, but is rather an outcome of it.

We take a similar approach in modeling the linkage between variety choice and household diversity outcomes for wheat in China. Level of risk, market development and access, as well as the physical characteristics of the growing environment, are all hypothesized to be factors that are considered by Chinese farmers in their choice of wheat varieties. We also assume that farmers do not explicitly recognize the maintenance of crop diversity as a factor in their variety choice decision. Consequently, the choice of varieties will largely determine diversity outcomes.

Crop Diversity

Crop diversity can be represented and measured in many different ways (Chapter 2). The choice of a diversity measure depends largely upon the context of its application and on available data. In this chapter, we use three measures of spatial diversity calculated at the household level, based on the number of named varieties cultivated in each household: (1) the total number of named varieties cultivated by the household (richness); (2) the Berger-Parker index of inverse dominance measuring the proportion of the area of a given variety out of total area planted; and (3) the Shannon index of evenness measuring the distribution of the proportions.

Data and Preliminary Statistical Analysis

Three significant wheat-producing provinces in China—Shandong, Shanxi, and Gansu—were selected for our household survey. These provinces represent the autumn-planted Yellow and Huai River Valleys facultative wheat zone, the autumn-planted northern winter wheat zone, and part of the spring-planted northwestern spring wheat zone and the autumn-planted northern winter wheat zone, respectively (He et al. 2001). Using a stratified sampling method, two counties with variation in topography and infrastructure development and market access were selected in each province. Five villages in each county were then chosen, also stratifying on topography and infrastructure along with market development. Approximately 10 randomly selected households were surveyed in each village. Key informants were also surveyed at the village, township, and county levels.

Wheat production plays an important role in all of the surveyed counties (Table 9.1). Area planted to wheat in 1997 accounted for 31% to 47% of all cropped land area in the surveyed counties of Gansu Province. The average area in 1997 for the entire province was 35%. The surveyed county with the lowest wheat-to-cropped area ratio, Yuci County in Shanxi Province, still maintained a ratio of wheat area to total cropped area of 22% in 1997, while the average over all counties for that year in Shanxi was approximately 25%. With 36% of area planted in wheat, the surveyed counties in Shandong reflected exactly the overall county average for that year. The range in value of the multiple cropping index at the village level, defined as the village's total cropped land area divided by the its cultivated land area, ranges from a high of 1.82 in Qinzhou County in Shandong Province to a low of 0.88 in Wenshui County in Shanxi Province.[1] Per capita cultivated area for all surveyed counties ranged from a low of 0.07 hectare per person in Qinzhou County of Shandong Province to a high of 0.18 hectare per person in Yuzhong County of Gansu Province. Data from the surveyed counties are on the whole slightly

[1] *Cropped area will be greater than cultivated area where more than one crop is sown during the cropping year.*

Table 9.1 Surveyed county and household characteristics, 1997.

		County			Farm household		
Province	*County*	*Wheat area (share of total cropped area) (%)*	*Per capita cultivated land area (ha/person)*	*Multiple cropping index*	*Household head education (years)*	*Per capita wheat consumption (kg/person)*	*Non-agricultural employment*[a] *(%)*
Shandong	Qingzhou	36	0.07	1.82	7.29	210	18
	Huimin	36	0.12	1.65	6.95	266	5
Shanxi	Yuci	22	0.09	1.02	7.47	167	22
	Wenshui	42	0.10	0.88	7.05	181	16
Gansu	Zhang	31	0.17	1.06	6.50	199	18
	Yuzhong	47	0.18	0.98	7.17	201	12

Sources: County data from county-level Statistical Bureau; authors' survey.
[a] For preceding year

higher than the national average for per capita cultivated land area (0.08 hectares per person) for 1997.

Households in Shandong Province, particularly Huimin County, also appear to be relatively more dependent on wheat as their main staple. Per capita consumption of wheat alone in several of the surveyed villages exceeded 200 kilograms. As a measure of comparison, average per capita consumption level of cereals (including wheat, rice, and maize) in rural China was 217 kg in 1997 (Huang and Li 2003).

About 15% of rural labor was employed full-time in the non-farm sector in our sampled households. This figure is consistent with the results from a nationally representative survey conducted by the Center for Chinese Agricultural Policy, Chinese Academy of Sciences (Rozelle et al. 2002), which found that full-time off-farm employment accounted for 16% of total rural labor in 1997, although there was substantial variation across provinces. The study by Rozelle et al. (2002) also shows that the share of off-farm employment in total rural labor increased to 22% and 36%, respectively, in 1997, if seasonal and part-time off-farm employment were included. In our surveyed counties, we found similar variation for full-time employment in the non-farm sector, ranging from 5% in Huimin County of Shandong Province to 22% in Yuci County of Shanxi Province. This employment structure provides sufficient variation to test the impacts of off-farm employment on wheat varietal diversity.

Observed Levels of Wheat Diversity

Regional differences in number of named varieties

Farmers' use of wheat varieties in recent years is shown in Table 9.2. The number of named varieties cultivated in 1997 ranges from a maximum of 17 in Zhang County, Gansu Province, to a minimum of 6 in Huimin County, Shandong Province. The number of named varieties in Gansu Province is also higher than that of the other two provinces for all years for which data were collected. One possible explanation for this difference is the cultivation of both winter wheat and spring wheat types in Gansu Province. None of the surveyed villages in Gansu Province was found to cultivate both winter and spring type wheats, although both types were found at the county level. Examining changes over time, the number of varieties grown in Wenshui County, Shanxi Province, increased very rapidly over a six year period. The number doubled between the first year of implementing the Household Responsibility System (Chapter 3) and 1992, and doubled again between 1992 and 1996.

In our sample, each household planted 1.5 varieties on average in 1997, with a range of one to four varieties. About 60% of households planted one variety only, 32% of households planted two varieties, and 10% of households planted 3-4 varieties (Table 9.3). Only in Huimin County, Shandong Province,

Table 9.2 Number of wheat varieties cultivated in surveyed counties.

Province	*County*	*1982*[a]	*1992*	*1996*	*1997*	*Average*
Shandong	Qingzhou	9	8	8	8	8
	Huimin	8	7	7	6	7
Shanxi	Yuci	10	9	8	9	9
	Wenshui	3	6	11	12	8
Gansu	Zhang	16	15	15	17	16
	Yuzhong	12	13	11	14	13

[a]The household responsibility system in most of the surveyed villages began in 1982, although the sample range is from 1980 to 1983. For simplification, we use the year 1982.

Source: Authors' survey.

and in both counties in Gansu Province, did the number of households cultivating more than one variety exceed the number of households growing only one variety. Of the 23 households cultivating more than two varieties, 18 were located in Gansu Province.

Table 9.3 Household cultivation of wheat varieties, 1997.

Number of varieties cultivated	*Yuci*	*Wenshui*	*Qingzhou*	*Huimin*	*Zhang*	*Yuzhong*	*Total # of households*
One variety	28	35	27	19	11	20	140
Two varieties	4	9	8	21	15	18	75
Three varieties	2	0	0	3	9	4	18
Four varieties	0	0	0	0	5	0	5
Sample size	**34**	**44**	**35**	**43**	**40**	**42**	**238**

Source: Authors' survey.

Determinants of wheat diversity

The availability of varieties and the characteristics embodied in each variety play a large role in determining household variety selection. Each variety is characterized by a recognizable package of traits whose relative importance will be perceived and weighted differently by different households. Household demand for variety traits can be seen as driving an implicit household demand for crop diversity. We now turn our attention to possible factors influencing variety choice and, by association, levels of wheat diversity.

The household's per capita cultivated land area appears to be correlated with crop diversity, as defined for this analysis based on the number of cultivated varieties (Table 9.4). It is possible that households with larger cultivated land area per capita can benefit by staggering their time and labor

Table 9.4 Analysis of factors influencing levels of wheat diversity.

Variable and value	*Average value*	*Genetic diversity indexes*		
		Number of named varieties (richness)	*Shannon (evenness)*	*Berger-Parker (abundance)*
Per capita cultivated area (ha/person)				
0.0133-0.0996	0.064	1.35	0.24	1.24
	(75)[a]	(0.61)[b]	(0.35)[b]	(0.44)[b]
0.0997-0.1889	0.141	1.55	0.29	1.29
	(104)	(0.75)	(0.38)	(0.47)
0.1900-0.7334	0.276	1.73	0.49	1.50
	(59)	(0.78)	(0.44)	(0.53)
Per capita wheat consumption (kg/person)				
45-149	113	1.41	0.24	1.27
	(69)	(0.60)	(0.37)	(0.47)
150-225	183	1.44	0.28	1.29
	(86)	(0.73)	(0.37)	(0.48)
226-450	306	1.72	0.42	1.42
	(83)	(0.79)	(0.41)	(0.49)
Proportion of non-farm employment				
0	0	1.61	0.38	1.42
	(96)	(0.77)	(0.43)	(0.55)
0.01-0.90	0.25	1.47	0.28	1.27
	(142)	(0.69)	(0.37)	(0.43)
Multiple cropping index of the village				
0.76-1.02	0.88	1.63	0.37	1.41
	(104)	(0.80)	(0.42)	(0.54)
1.03-1.29	1.15	1.55	0.30	1.31
	(51)	(0.81)	(0.41)	(0.48)
1.30-4.39	2.17	1.40	0.26	1.24
	(83)	(0.54)	(0.35)	(0.40)

[a] Sample total of the specific grouped data
[b] Standard error of the corresponding wheat diversity of each group
Source: Authors' survey

resources through the cultivation of more than one variety, perhaps ensuring differing maturity periods. The average number of cultivated varieties increases from 1.35 to 1.73, when per capita cultivated land increases from 0.064 hectares to 0.276 hectares. Within the same range of per capita cultivated land, the average Shannon evenness index increases 104% from 0.24 to 0.49, while the average Berger-Parker inverse dominance index increases 21% from 1.24 to 1.50.

A strong relationship also appears to exist between per capita wheat consumption and wheat diversity. Higher levels of household wheat consumption suggest a relatively greater role of wheat in the household's livelihood. To satisfy its full range of consumption needs, a household may decide to cultivate the set of varieties that can best meet the requirements for different uses (home consumption, markets sales, etc.). The average number of named varieties increases 22% and the average Shannon index increases 75%, when per capita wheat consumption rises from 113 kg per person to 306 kg per person (Table 9.4). Similarly, for the same range in per capita wheat consumption, the average Berger-Parker index increases 12%.

Due in large part to rapid economic development across China, the opportunity cost of agricultural labor has been increasing (Chapter 3). The increase in off-farm employment has significantly improved farmers' incomes. The higher the opportunity cost of rural labor, the more unwilling the household may be to allocate valuable labor resources for sowing, managing or harvesting multiple crop varieties. Income from non-farm sources can also enhance a household's risk-bearing ability and reduce its need to spread risk through multiple varieties of the same crop. The grouped data in Table 9.4 show that measures of diversity for households whose members do not work off-farm are all higher than those of the households with members employed off-farm by a difference ranging from 9% to 26%.

The multiple cropping index (MCI) of the village also exhibits an inverse relationship to individual crop diversity. Classifying villages into three groups by MCI values, the diversity indices (number of named varieties, Shannon, and Berger-Parker) of the group of villages with the lowest MCI values are 14%, 30%, and 12% higher, respectively, than the group of villages with the highest MCI values.

Econometric Model and Estimation Results

Econometric model

Household decisions regarding production activities are affected not only by individual household characteristics, but also by the type of risks the households face, their ability to cope with risk, their production environment, and the market environment. Based on existing literature, we set up the following reduced form econometric model:

$$D_{hjk} = f\,(H_{hjk}, R_{hjk}, Q_{hjk}, S_{hj}, Z_{hj}, P_{hk}, G_{hjk}) + e_{hjk} \qquad (9.1)$$

where, h (h=1, 2... , 238) represents individual farm households, j (j=1, 2,... , 30) represents villages, k (k=1, 2, 3) represents provinces, and D represents the calculated crop diversity indices. H is a vector of household characteristics, including household size, age of the head of household, and education level of the head of household. Education and age of the household head can reflect attitudes towards existing technology, perceived risk, and access to information. If older farmers are indeed more likely to be risk averse (Meng et al. 1998; Gould et al. 1989; Bultena and Hoiberg 1983), we would expect to observe more risk diversification measures taken, such as the cultivation of multiple varieties with differing traits or management requirements. If farmers with higher levels of education are more willing or able to deal with perceived risk (Moscardi and de Janvry 1977; Nelson and Harris 1978), we would expect fewer such measures to be taken; farmers with higher levels of education may also have better access to information or technologies that could replace the need for multiple varieties.

R is a vector of variables reflecting the ability of the household to bear risk. We include the multiple cropping index to represent diversification in agricultural activities at the village level, to avoid the strong endogeneity problem associated with a similar index at the household level, as well as the proportion of time that household labor is involved in non-agricultural activities. *Q* represents the effect of market development on crop diversity and includes variables such as per capita wheat consumption in the household (kg/person), the degree of commercialization in the household defined as the ratio of the quantity sold to the total household wheat production, and local marketing and transportation conditions.

S represents plot level characteristics, including the area and quality of cultivated land in the household, to account for the ways in which differences in production environment influence farmers' choices of variety and crop diversity. It includes a Shannon index calculated for diversity in soil quality (high fertility, average fertility, below average fertility, very low fertility); the ratio of sloped field area to the total field area; average distance from the plot to the household dwelling; and per capita cultivated area (ha/person). *Z* represents local variety supply using the number of named varieties in a village (the number of named varieties per approximately 1,000 ha), while *P* includes a variable for wheat type (winter wheat or spring wheat) and a provincial dummy variable. *G* reflects regional differences with a set of regional indicator variables. Finally, we define *e* as a random error term. The hypothesized impacts of each factor on wheat diversity are summarized in Table 9.5.

The wheat diversity indices used in this chapter all have defined range limits. The richness index ($D \geq 0$) and Berger-Parker relative inverse dominance index ($D \geq 1$) are bounded at on one end, while the Shannon evenness index ($0 \leq D \leq \frac{n}{e}$) is bounded at both ends. As a result, the statistical results by OLS model may be biased. We therefore utilize the Tobit model and maximum likelihood methods to estimate the wheat diversity model specified in Equation (9.1). In the ordinary form of the Tobit model (for the sake of convenience we drop the subscripts),

$$\begin{aligned} D &= f(H, R, Q, S, Z, P, G) + e \quad (9.2) \\ &= \beta' X + e \end{aligned}$$

$$D^* = \begin{cases} d_1 & \text{if } D^* \leq d_1 \\ D & d_1 \langle D^* \langle d_2 \\ d_2 & \text{if } D^* \geq d_2 \end{cases} \quad (9.3)$$

Table 9.5 Hypothesized impacts of various factors on wheat diversity.

Independent variables	*Expected sign*
Household characteristics (H)	
Household size	−/+
Age of household head	+
Educational level of household head	−
Party cadre	+
Variables related to risk (R)	
Village's multiple cropping index	−
Non-agricultural employment of family members	−
Wheat consumption and market (Q)	
Per capita wheat consumption	+
Commercialization level	−/+
Distance from town	+
Cultivated land conditions (S)	
Per capita cultivated area	+
Field–house distance	−
Number of household plots	+
Proportion of sloping fields	+
Soil diversity index	+
Variety supply (Z)	
Number of wheat varieties available in previous year	+

where β is the vector of coefficients to be estimated, X is the vector of independent variables, d_1 is the lower limit of the wheat diversity index, and d_2 is the upper limit of the wheat diversity index. The likelihood equation of the Tobit model is as follows:

$$L(\beta,\alpha|D,X,d_1,d_2)=\prod_{D^*=d_1}\Phi\left(\frac{d_1-\beta' X}{\alpha}\right)\prod_{D^*=D}\frac{1}{\alpha}\phi\left(\frac{D-\beta' X}{\alpha}\right)\prod_{D^*=d_2}\left[1-\Phi\left(\frac{d_2-\beta' X}{\alpha}\right)\right] \tag{9.4}$$

where $\Phi(.)$ is the cumulative distribution function, $\phi(.)$ is the density function, and α is the standard error.

Estimation results

In the previous section, we discussed some simple observed relationships between wheat diversity and certain explanatory variables that we hypothesize to be important. However, as Table 9.5 shows, wheat diversity is expected to be simultaneously affected by a set of variables, and attributing influence requires empirical estimation of the model developed in the previous section. Table 9.6 reports the estimation results of household wheat diversity based on the survey sample of 238 households.

Table 9.6 Estimation results of household wheat diversity: Tobit model.

	No. of named varieties	*Shannon index*	*Berger-Parker index*
Household member characteristics (H)			
Household size	–0.101	–0.052	–0.062
	(–1.09)	(–1.08)	(–1.04)
Age of household head	–0.015	–0.002	–0.005
	(–1.46)	(–0.54)	(–0.71)
Household head educational	0.005	–0.001	–0.007
	(0.16)	(–0.04)	(0.30)
Party cadre	0.975	0.65	0.624
	(1.88)*	(2.49)**	(1.91)*
Variables related to risk (R)			
Multiple cropping index	–0.484	–0.159	–0.205
	(–2.60)***	(–1.76)*	(–1.79)*
Non-agricultural employment of family members	–0.954	–0.526	–0.728
	(–1.63)	(–1.73)*	(–1.92)*
Consumption/market (Q)			
Per capita wheat consumption	0.002	0.001	0.001
	(1.77)*	(1.92)*	(1.60)
Commercialization level	1.164	0.095	0.118
	(2.17)**	(0.35)	(0.35)
Distance from town	–0.010	–0.002	–0.006
	(–0.41)	(–0.14)	(–0.42)
Cultivated land (S)			
Per capita cultivated area	–0.738	1.012	1.063
	(–0.58)	(1.65)*	(1.38)
Field–house distance	0.174	–0.054	–0.095
	(1.09)	(–0.66)	(–0.92)
Number of household plots	0.209	0.072	0.072
	(3.44)***	(2.32)**	(1.88)*
Proportion of sloping fields	0.918	0.278	0.428
	(2.34)**	(1.40)	(1.74)*
Soil diversity index	0.411	0.166	0.226
	(1.42)	(1.11)	(1.21)
Variety supply (Z)			
Number of wheat varieties available in previous year	2.186	1.3	1.166
	(3.04)***	(3.62)***	(2.64)***
Wheat type (P)			
Winter wheat	0.532	0.189	0.243
	(1.33)	(0.93)	(0.97)

Region (G)			
Shandong	−1.400	−0.792	−0.892
	(−3.04)***	(−3.48)***	(−3.16)***
Shanxi	−1.262	−0.753	−0.956
	(−3.67)***	(−4.30)***	(−4.38)***
Intercept	0.877	−0.083	1.202
		(−0.21)	(2.40)**
Sample size	238	238	238

Z test significant at 10% (*), 5% (**), and 1% (***) levels

The results indicate that numerous factors are significant in determining levels of wheat diversity at the household level.

1. *Household characteristics do not have a consistently significant effect on diversity* Most household characteristics, including age and educational level of the household head, were not significant in most estimation results. However, the effect of the village cadre variable is significant. A household containing a village cadre maintains a higher level of crop diversity than other households, all other factors being equal. Village cadres are a special group in Chinese rural areas. They are often the first to receive information about and to adopt new varieties and thus also play an important role in the dissemination of new technologies. Households with village cadres are more likely to experiment with newly released varieties.
2. *Farmer ability to bear risk has a significant negative effect on diversity* The village level MCI has a significant and negative effect on crop diversity. There are two possible explanations. First, a high MCI suggests a more intense use of available land with a higher corresponding level of restrictions on crop duration. Opportunities to choose from a wide range of wheat varieties may consequently be more limited. Secondly, an increase in the MCI implies greater opportunities for diversification across all agricultural activities. These opportunities mitigate the pressure to depend on the use of multiple varieties of one crop to disperse risk and thus may decrease observed levels of wheat diversity.

   The negative and significant coefficient of the proportion of household labor time involved in non-agricultural employment implies that an increase in non-agricultural opportunities enhances farmers' risk bearing-ability and reduces the need to use multiple varieties as a risk-spreading measure. Crop diversity, as defined by the number of named varieties, will consequently decrease. In addition, the proportion of non-agricultural employment represents the opportunity cost of labor. If more labor is required in cultivating multiple varieties, households will be less likely to accept the rises in cost as their labor becomes increasingly valuable.

3. *Household wheat consumption and commercialization have significant effects on diversity* In the specifications using the number of named varieties and the Shannon evenness index as the dependent variable, coefficients for per capita wheat consumption are significant. A higher level of per capita consumption indicates a higher level of household dependence on wheat, and may also imply more numerous requirements for end use and thus for a range of available traits. Moreover, in the results of the specification using the number of named varieties, the household's wheat commercialization level has a significant and positive effect on crop diversity. An increase in the percentage of household wheat sold thus increases the incentive for farmers to cultivate multiple wheat varieties. Households may consciously select varieties with certain traits, such as high yields, for marketing purposes, while cultivating others with better consumption traits for use in the household.
4. *Most characteristics related to household land significantly affect diversity* The coefficient of per capita cultivated land area is positive and significant in the Shannon index specification, suggesting that per capita cultivated land area is an important determinant of household wheat diversity. If the amount of household labor resources is fixed, households with more land may need to plant different varieties to utilize their limited labor resources most efficiently. The coefficient on the number of household plots is also positive and significant. The more plots cultivated by a household, the higher the level of wheat diversity in the household. In the specifications using the number of named varieties and the Berger-Parker index as the dependent variable, the proportion of sloping fields is positive and significant. Sloping fields tend to be affected more by soil and water erosion than level land, and in many circumstances can also be less accessible.
5. *Local variety supply is an important factor in diversity* The results from the estimation of all specifications show that local variety supply has the most significant and positive effects on wheat diversity. As expected, as more wheat varieties become available in the local markets, farmers have more choices to meet their diversified demands in wheat production, consumption, and marketing.

Conclusions and Policy Implications

This study analyzes factors affecting levels of wheat diversity measured at the household level using primary survey data from 238 farm households in Shandong, Shanxi, and Gansu Provinces in China. We calculate three indices of spatial diversity based on information on the number of named varieties cultivated in each household. Our findings indicate that enhancing households' ability to bear risk, improving household plot level production characteristics, and raising the opportunity cost of labor will have a negative effect on household wheat diversity. Even when the other explanatory variables are taken into account, Gansu Province still appears to have the highest diversity levels. This result may reflect a higher level of general agro-ecological variation in the province relative to either Shanxi or Shandong Provinces. The findings may also be related to the level of overall development in the provinces, as reflected by the provincial dummy variables, which seems to have a detrimental effect on household diversity levels. Higher levels of per capita wheat consumption in the household and more options in terms of locally available and locally adapted wheat varieties, on the other hand, will help to promote diversity at the household level.

These results provide some directions for strategies to improve crop diversity at the household level that will have impacts on diversity at more aggregate levels. However, we emphasize two points in considering possible implications of this research. First, the diversity variables used in this study are based on varieties named during the household surveys. The use of named varieties does not specifically incorporate information on similarities in traits among varieties or shared parentage, all of which could affect diversity outcomes. A comparison of findings using diversity indices calculated with morphological characteristics and/or pedigree data would therefore be informative. Furthermore, diversity measures based on named varieties do not take into account variation found within a given variety. While the extent of the possible range of variation for wheat, a self-pollinating crop, is limited, the use of farmer-saved seed over multiple years was widely observed and could possibly result in unexpected levels of variation within a given variety or intra-varietal diversity. Households cultivating only one named variety may consequently maintain a higher level of diversity than would be otherwise evident.

Second, with very few exceptions, the wheat varieties cultivated in our survey area are all modern varieties; that is, varieties that have been improved through breeding efforts. Our results thus provide an interesting comparison with previous work on household diversity outcomes that have taken place in centers of crop diversity or domestication where traditional varieties are still widely cultivated (Smale 2006; Van Dusen and Taylor 2005; Smale et al. 2001; Meng et al. 1998; Brush et al. 1992). Farmers in many of these regions have been recognized as an important force in maintaining diversity levels in the context of *in situ* conservation of crop genetic resources. Given the complex interaction of household production and consumption incentives for choosing varieties, the prospects for maintaining diversity in these centers are often encouraging.

However, while per capita consumption of wheat remains a significant factor influencing variety choice for Chinese households, it is less clear that a similarly complex interaction of incentives comes into play in the variety choice decision, particularly with respect to the role of cultural and consumption traits associated with specific wheat varieties. The prospects of crop diversity at the household level in areas of improved varieties thus may not be as promising outside of centers of diversity. Given the decrease or absence of strong cultural or consumption reasons to cultivate multiple varieties, as a household's ability to diversify risk through other means increases or to smooth out the effects of household-level agro-climatic differences, the household's need to rely on crop diversity will also decrease, and it is likely that they may increasingly de-emphasize wheat in their household portfolios.

External policy influences may also have a negative effect on household diversity outcomes. To improve the ecological environment, China has implemented a land conversion program in certain regions where fields exceeding a pre-determined slope angle must be converted to forest. This policy, while beneficial in aspects of soil erosion, may have a negative effect on levels of household crop diversity, and this type of trade-off reflects potential dilemmas for policy makers.

Diversity in improved varieties, while not necessarily relevant to in situ conservation in the same way, has been recognized as significant in maintaining productivity levels (Chapters 10 and 11; Jin et al. 2002). Based on the research results, we propose two primary areas that merit additional attention if the existence of crop diversity at the household level is viewed as important. First, variety options for households should be increased through the development of new varieties and/or through the improvement of supply channels. A larger pool of choices for the household can often have a positive effect on diversity levels. Second, crop breeding objectives should continue to be broadened beyond the traditional focus on yield increases. Research priorities should continue to expand to develop varieties for multiple objectives (e.g., consumption quality and processing requirements) that meet the needs of farmers and markets. In particular, given rising income trends and negative income elasticity of demand for wheat across population groups in China, a focus on the development of wheat varieties of specific high quality attributes, may be increasingly warranted. Research priorities should also be placed on the utilization of a diverse set of genetic materials. If households in the future elect to cultivate fewer varieties, the importance of scientifically-maintained diversity within a given variety will become increasingly important. Additional exploration of diversity linkages between household and community levels is also required to understand implications of decreased diversity at household levels on aggregate level diversity.

REFERENCES

Ames, G.C. and D.W. Reid. 1993. Risk analysis of new maize technology in Zaire: A portfolio approach. *Agricultural Economics* 9: 203-214.

Bellon, M. and J.E. Taylor. 1993. Folk soil taxonomy and the partial adoption of new seed varieties. *Economic Development and Cultural Change* 41: 763-786.

Brush, S.B. 2004. *Farmers' Bounty: Locating Crop Diversity in the Contemporary World*. New Haven: Yale University Press.

Brush, S.B., J.E. Taylor, and M.R. Bellon. 1992. Technology adoption and biological diversity in Andean potato agriculture. *Journal of Development Economics* 39: 365-387.

Bultena, G.L. and E.O. Hoiberg. 1983. Factors affecting farmers' adoption of conservation tillage. *Journal of Soil and Water Conservation* 38: 281-284.

de Janvry, A., M. Fafchamps, and E. Sadoulet. 1991. Peasant household behavior with missing markets: Some paradoxes explained. *The Economic Journal* 101: 1400-1417.

Fafchamps, M. 1992. Cash crop production, food price volatility, and rural market integration in the third world. *American Journal of Agricultural Economics* 74: 90-99.

Feder, G. 1980. Farm size, risk aversion, and the adoption of new technology under uncertainty. *Oxford Economic Papers* 32: 263-284.

Just, R.E. and D. Zilberman. 1983. Stochastic structure, farm size and technology adoption in developing agriculture. *Oxford Economic Papers* 35: 307-328.

Gould, B.W., W.E. Saupe, and R.M. Klemme. 1989. Conservation tillage: The role of farm and operator characteristics and the perception of erosion. *Land Economics* 65: 167-182.

He, Z.H., S. Rajaram, Z.Y. Xin, and G.Z. Huang (eds.). 2001. *A History of Wheat Breeding in China*. Mexico, D.F.: CIMMYT.

Huang, J. and N. Li. 2003. China's Agricultural Policy Simulation Model (CAPSiM). Center for Chinese Agricultural Policy, Chinese Academy of Sciences.

Jin, S., J. Huang, R. Hu, and S. Rozelle. 2002. The creation and spread of technology and total factor productivity in China's agriculture. *American Journal of Agricultural Economics* 84: 916-930.

Lee, D.R. 2005. Agricultural Sustainability and Technology Adoption: Issues and policies for developing countries. *American Journal of Agricultural Economics* 87(5): 1325-134.

Meng, E., J.E. Taylor, and S.B. Brush. 1998. Implications for the conservation of wheat landraces in Turkey from a household model of varietal choice. In *Farmers, Gene Banks and Crop Breeding: Economic Analyses of Diversity in Wheat, Maize, and Rice*. M. Smale (ed.), Boston: Kluwer Academic Publishers.

Morduch. J. 1995. Income smoothing and consumption smoothing. *Journal of Economics* 9: 103-114.

Moscardi, E. and A. de Janvry. 1977. Attitudes toward risk among peasants: An econometric approach. *American Journal of Agricultural Economics* 59: 710-716.

Nelson, A.G. and T.D. Harris. 1978. Designing an instructional package: The use of probabilities in farm decision making. *American Journal of Agricultural Economics* 60: 993-997.

Rozelle, S., A. deBrauw, L. Zhang, J. Huang, and Y. Zhang. 2002. The evolution of China's rural labor markets during the Reforms: Rapid, accelerating, transforming. *Journal of Comparative Economics* 30: 329-353.

Smale, M. (ed.). 2006. *Valuing Crop Diversity: On-farm Genetic Resources and Economic Change*. Wallingford, UK: CABI Publishing.

Smale, M., M.R. Bellon, and J.A. Aguirre Gomez. 2001. Maize diversity, variety attributes, and farmers' choices in Southeastern Guanajuato. *Economic Development and Cultural Change* 50: 201-225.

Van Dusen, M.E. and J.E. Taylor. 2005. Missing markets and crop diversity: Evidence from Mexico. *Environment and Development Economics* 10: 513-531.

Chapter 10

Wheat Diversity and Total Factor Productivity in China

S. Jin, E. Meng, S. Rozelle, R. Hu, and J.K. Huang[1]

ABSTRACT

An improved understanding of the relationship between crop diversity and productivity is important, particularly if appropriate policies for diversity are to be identified and implemented. The impact of wheat diversity on the productivity of wheat in China is examined using total factor productivity (TFP) and an instrumental variable approach. TFP in the seven key wheat-producing provinces in China from 1982 to 1995 shows significant, though variable, growth for all provinces in that period. Analysis of the causes of TFP growth tested alternative taxonomies of diversity (named varieties and morphological groups) and three measures of diversity (richness, inverse dominance, and evenness). This analysis reveals that diversity significantly affects TFP and that the results are consistent across taxonomies and measures of diversity.

With relatively little information about the net impact of crop diversity on productivity, policy makers in developing countries may be reluctant to invest scarce financial resources for the conservation and utilization of crop diversity, particularly if the primary benefits are believed to address scientific or environmental concerns in the future, rather than ensuring the immediate food security needs for growing populations. Political economy realities will also discount welfare-increasing benefits for small-scale farmers in marginal or isolated communities.

In this chapter we analyze the impact of wheat diversity on the productivity of wheat at the national level in China during 1982-95. To accomplish this, we examine the effects of crop diversity on total factor productivity (TFP), a measure of technical efficiency that is commonly used to gauge sectoral performance, utilizing a subset of the spatial diversity indices

[1] *This chapter draws on Jin et al. 2002.*

previously discussed in Chapter 6. The scope of the analysis is seven key, wheat-producing provinces: Hebei, Shandong, Henan, Shanxi, Jiangsu, Anhui, and Sichuan. All except Shanxi were in the top six wheat producing provinces each year during the study period, and together the seven provinces accounted for an average of 63% of China's sown area and 71% of its output over the study period.

Analysis of Wheat Productivity in Post-reform China

Differences in the estimates of China's TFP calculated by Tang and Stone (1980) and Wiens (1982) initiated a debate on the success of pre-reform agriculture. In the major work documenting TFP growth in the reform era, Wen (1993) confirmed the efficiency analyses of McMillan et al. (1989) and Lin (1992), concluding that rapid TFP growth had at least partially fueled the rural economy's miracle growth in the early 1980s. However, because Wen's analysis covered only a period through 1990, it created the erroneous impression that the agricultural sector was in trouble due to the stagnation of aggregate TFP growth after 1985. The resulting conclusion that productivity could have fallen in the late 1980s was later put in doubt due to the continued growth of output in the agricultural sector at over 5% per year.

Contributing to the debates and uncertainty revolving around previous productivity studies are poor data and ad hoc weights. Data sources are numerous and not necessarily consistent, and all researchers warn of the poor quality of many of the input and output series. Stone and Rozelle (1995) caution that the trends of all pre-reform TFP estimates depend heavily upon the nature of the assumed factor proportions used to aggregate inputs. Without a means of determining the most appropriate set of weights, Wen (1993) utilized sensitivity analysis, updating aggregate TFP until the early 1990s with all three sets of weights devised by earlier analysts.

Methodology for TFP Measures

The methodology for calculating our TFP measures is described in detail in Jin et al. (2002). Conceptually, TFP is the examination of an index of output changes relative to an index of input changes. Changes in total output not accounted for by the changes in total inputs are attributed to technical advances. For a homogenous commodity, TFP can be computed as a ratio of output to an aggregated index of inputs used in the production of the output. In our study, a Turnquist-Theil index is applied to compute wheat TFP by province over time. Expressed in logarithmic form, the Tornquist-Theil TFP index is defined as:

$$\ln (TFP_t/TFP_{t-1}) = \ln (Q_t/Q_{t-1}) - \tfrac{1}{2} S_j (S_{jt} + S_{jt-1}) \ln (X_{jt} / X_{jt-1}), \quad (10.1)$$

where Q is wheat production (output); S_{jt} is the share of input j in the total cost of wheat production; X_j is input j used in the production of wheat; and t indexes time (years).

Setting TFP in the base year to 100 and accumulating the changes over time based on Equation (10.1) provides a time series of TFP indices for each province. The Tornquist-Theil index is exact for the linear homogeneous translog production function (Diewert 1976) and superlative under very general production structures, i.e., non-homogeneous and non-constant returns to scale (Caves et al. 1982). It also provides consistent aggregation of inputs and outputs under the assumptions of competitive behavior, constant returns to scale, Hicks-neutral technical change, and input and output separability. Because current factor prices are used in the construction of the weights in aggregating the input index, quality improvements in inputs are also incorporated (Capalbo and Vo 1988). A similar approach is used by Rosegrant and Evenson (1992) in their agricultural productivity analysis for South Asia. Our methodological approach for calculating TFP is also similar to that of Rosegrant and Evenson (1992) and Fan (1997) in utilizing standard Divisia index methods.

Data Sources for TFP Measures

Data used in calculating TFP measures are also described in detail in Jin et al. (2002). We address some of the difficulties faced by previous researchers by utilizing a detailed data set collected by the State Price Bureau (SPB) over the past 20 years on the costs of production for all of China's major crops. The sampling framework covers more than 20,000 households and the data include information on quantities and total expenditures for all major inputs, as well as expenditures for a large number of miscellaneous costs. Data on output and total revenues earned from the crop are also reported for each household. The household-level data set is supplemented by provincial surveys conducted by the SPB that provide information on unit labor costs, reflecting the opportunity cost of the daily wage foregone by farmers.

A key addition to the data is a set of land rental rates collected by the authors during a 1995 survey of 230 villages in 7 provinces in China. Estimates of the average per hectare rental rate that farmers were willing to pay for cropping were obtained. These rates were elicited net of all other payments that are often associated with land transfer transactions in China (e.g. taxes), but which are picked up as part of the regular cost-of-production survey.

TFP Trends in Post-reform China

When aggregating across the seven provinces for an all-China wheat index, our TFP measure traces out a contour for the 1970s and 1980s similar to those found by Wen (1993) (Figure 10.1). Pairwise correlation coefficients between our measure and each of the measures based on Wen's fixed weights all exceed 0.85. During the early reform period of the early-mid 1980s, TFP for wheat rose more than 60%, an increase undoubtedly caused at least in part by

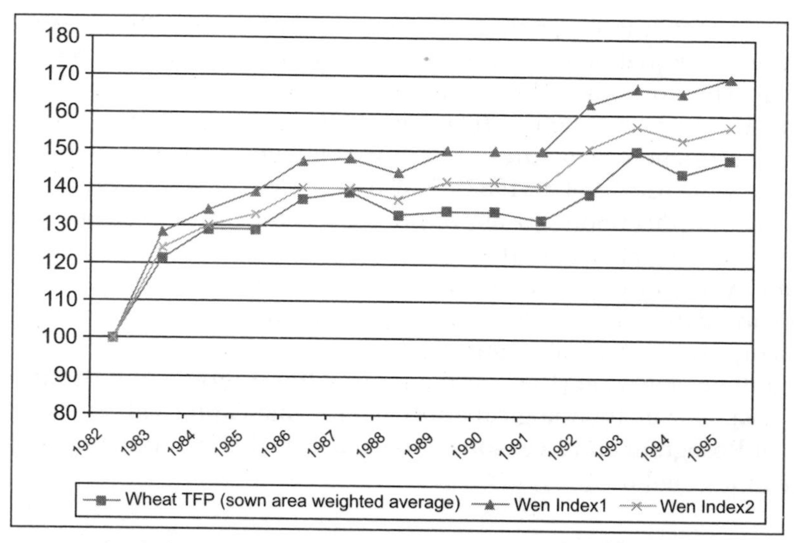

Figure 10.1 Total factor productivity indices, 1982-95[2].

incentives from the Household Responsibility System (Lin 1992) described in Chapter 4. Huang and Rozelle (1996), however, show that public investment in research and irrigation also contributed heavily to supply increases during this period. We would therefore expect these factors also to contribute to TFP growth.

TFP for wheat leveled off in the mid- to late-1980s in a trend that affected the entire agricultural sector, including other major cereal crops such as rice and maize. Proposed hypotheses for the stagnation in productivity include commodity pricing policies, land rights issues, and input availability (Jin et al. 2002). Decreasing levels of investment in research, water, and other public services may have also contributed to a slowdown in the turnover of new varieties, a fall in the release of varieties with higher yield potential, or a decline in the availability of varieties to meet new production, marketing and environmental challenges (Jin et al. 1999). Huang and Rozelle (1995) demonstrate that environmental degradation was also a factor in slowed output growth during this period.

Wheat TFP resumed its positive growth in the 1990s; wheat productivity rose significantly between 1990 and 1995, making it the best performing of the major cereal crops (Jin et al. 2002). Several factors could account for the resumption of growth. Reforms resulting in greater market liberalization during this period may have allowed producers to move into crops in which they had a comparative advantage. Leaders also refocused their efforts on investments in the research system (Rozelle et al. 1997), although the level of

[2] *Index 1 is calculated using a Tornquist-Divisia formula; Indices 2 and 3 are fixed coefficient indices from Wen.*

activity differs sharply by province. With the resurgence of growth, the average annual growth rates of TFP range between 1.8% and 2.9%, depending on the measure used.

Visually, TFP can be considered to be the gap between the output and input index trends in Figure 10.2. As the TFP analysis conducted in this chapter focuses specifically on wheat, the output index is calculated only for wheat. Data on wheat production inputs used in the computation for wheat TFP include sown area, labor, seed, fertilizer, pesticide, farm plastic film, animal traction, machinery and equipment, and other material inputs.

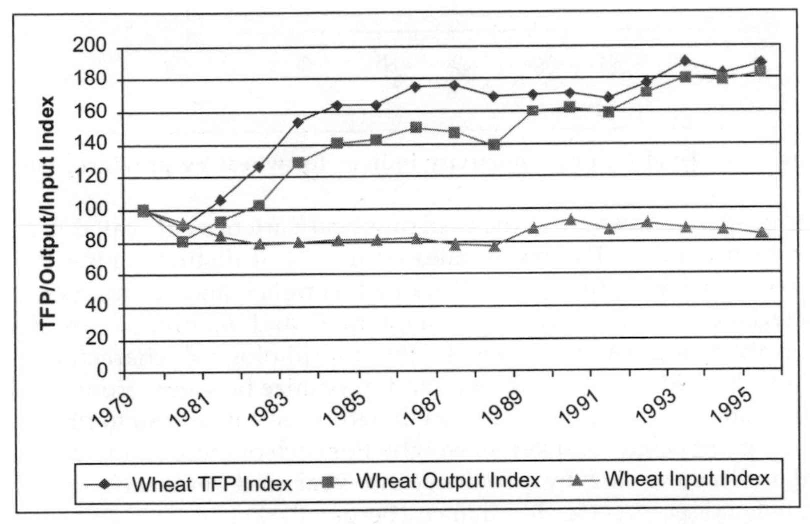

Figure 10.2 Output and input indices for wheat in China, 1982-95.

TFP growth across provinces varies sharply in both levels and trends. Figure 10.3 shows TFP trends in wheat by province during 1982-95. At 7.8%, the annual growth rate in Hebei Province was the highest, followed by 4.7% in Shandong and 3.5% in Henan. The higher levels and generally increasing trends in TFP observed in Hebei and Shandong Provinces are clearly not matched in Sichuan, Jiangsu, or Anhui Provinces, which had annual growth rates of 1.8%, 1.5%, and 0.5%, respectively. Hebei's rising TFP even during 1985-90 indicates that it does not appear to have experienced the stagnation in overall agricultural TFP during the same period. Wheat TFP in Shandong decreased slightly during this period but was able to recover subsequently. All other provinces also experienced stagnation or declines in TFP during 1985-90 that generally continued through to 1995.

Wheat Spatial Diversity

Spatial diversity is the variation within a given geographical area. To examine the impact of spatial diversity, we use six indices that capture different

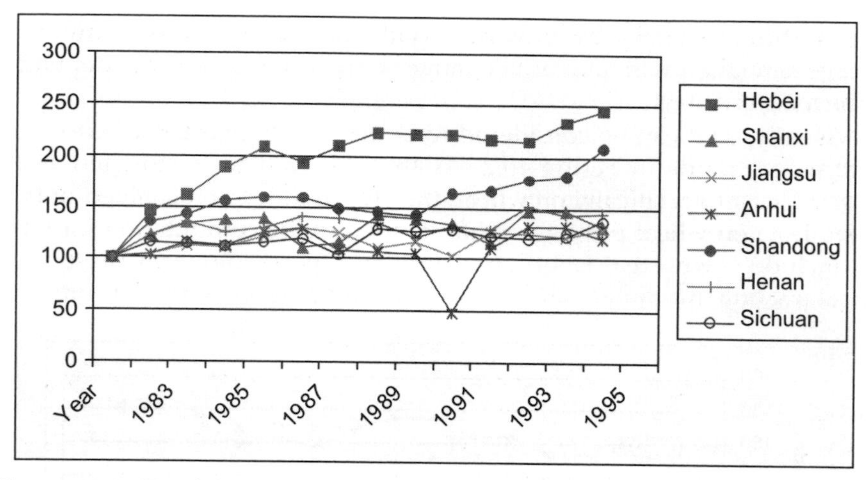

Figure 10.3 Total factor productivity indices for wheat by province, 1982-95.

aspects of spatial diversity in the pool of wheat varieties cultivated in each of the seven provinces. The taxonomies or means of distinguishing the crop population that we used were (1) named varieties and (2) morphological characteristics. As described in Chapters 2 and 6, groups are formed statistically using variation across the morphological characteristics to minimize the within-group variance and maximize between-group variance. Variety characteristics used were growth habit, resistance to stem rust, time to maturity, plant height, and kernel weight. For each of these taxonomies, three spatial indices varying by province and year and representing richness (Margalef index), inverse dominance (Berger-Parker index), and evenness (Shannon index) are calculated, as described in Chapter 2. Data for the construction of the diversity indices are drawn from a database of varieties, sown areas, and traits for major varieties compiled from government publications (MOA 1986-97), databases, and library materials, as well as communications with breeders in the seven selected provinces.[3]

Spatial diversity measured in these three ways varies across provinces in both levels and changes over time. As described in more detail in Chapter 6, regardless of the taxonomy used and despite year to year variation, levels in all three representations of spatial diversity (richness, inverse dominance, and evenness) are generally higher over the study period in Hebei, Shandong, and Henan Provinces, three of China's most important wheat growing areas (see Figures 6.7 and 6.8 in Chapter 6). A comparison across the richness, dominance, and evenness indices calculated using the data set of named

[3]*Only "major" varieties are counted in this calculation. A "major" variety in our sample is any variety that has a cultivated area of at least 6,667 hectares (10,000* mu) *in a province. Our database thus does not have full coverage of all varieties in each study province. However, the proportion of area covered by "major" varieties exceeds 90% in each province.*

varieties shows that, in general, wheat in Hebei and Henan is the most diverse throughout the study period, while Jiangsu and Sichuan Provinces generally rank among the least diverse for wheat. Diversity levels in Shanxi fall sharply towards the latter part of the study period. A similar situation holds upon examination of the set of diversity indices based on morphological groups. Sichuan Province stands out as the least diverse regardless of the type of spatial diversity examined.

TFP and Spatial Diversity

A visual observation of trends in total factor productivity and selected indices of spatial diversity over time for Hebei and Jiangsu Provinces shows interesting patterns (Figure 10.4). Hebei, the province with the highest TFP growth rate, also exhibits high levels and overall positive trends in spatial diversity as measured by selected indices (both morphological groups and named varieties). Provinces with slower TFP growth, such as Jiangsu, also had lower levels and either slower rates of increase or stagnant measures of diversity.

The above describes an empirical relationship between TFP and spatial diversity, but the causal relationship is complicated and multi-dimensional. Diversity can affect productivity both negatively and positively. On the positive side, a greater amount of diversity can provide farmers in heterogeneous agro-climatic environments with varieties better adapted for a range of soil and climatic conditions. A diverse set of varieties and/or varietal characteristics can also help slow or prevent the evolution and spread of new strains of pathogens or pests. In an area with poorly functioning markets, farmers with access to greater spatial diversity can cultivate a number of different varieties with different planting times and maturity characteristics, contributing to the more efficient use of fixed family labor and increased productivity.

A more diverse basket of varieties can also reduce total factor productivity. For example, if, lacking other means, farmers chose to cultivate a number of different varieties as a purely risk-reducing measure, productivity could decrease. Potential gains from specialization might also be reduced.

ANALYZING THE RELATIONSHIP BETWEEN TFP AND DIVERSITY

Model Specification

Wheat TFP in China is likely to be affected by many factors in addition to levels of spatial diversity. These include changes in R&D investment, technological advances, institutional reforms, infrastructure development, and improvements to human capital. The need to incorporate changes in resource quality into TFP analyses has also been widely discussed (see Murgai 2001; Ali and Byerlee 2002). Whether human capital endowments

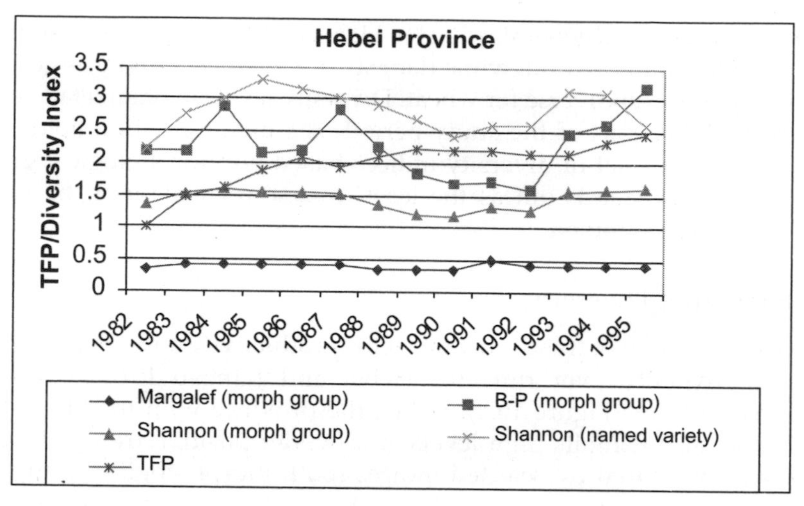

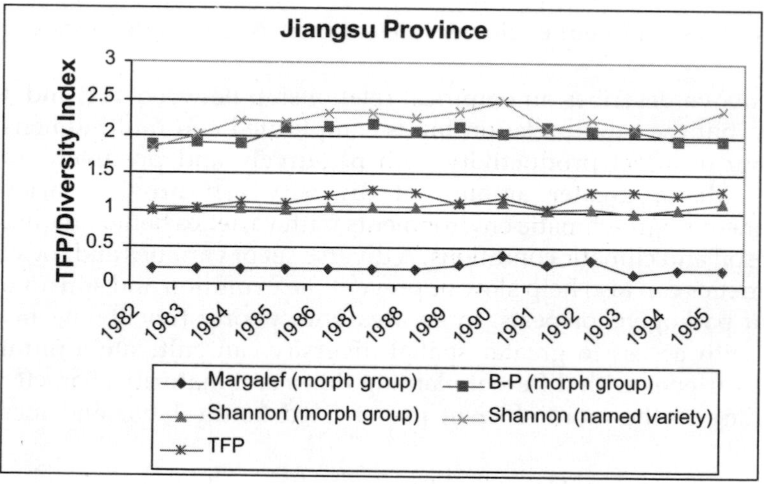

Figure 10.4 TFP and selected diversity indices, Hebei and Jiangsu provinces, 1982-95.

should be included in the determinants of TFP depends on how the measure is generated. For example, if current wages are used as a weight for labor input (as in this chapter), human capital is commonly assumed to have been accounted for. Our framework for explaining TFP changes over time can be specified as:

$$\text{TFP} = f\,(\textit{Spatial diversity, Technology, Infrastructure, Institutional Reforms, Z}) \qquad (10.2)$$

where Z is a vector of control variables whose elements represent weather, agro-climatic zones, and some fixed but unobserved factors that differ across regions. In most countries, technology and infrastructure are believed to be the major factors that drive long-term TFP growth (Rosegrant and Evenson 1992).

Most other determinants contribute either to short-term fluctuations (i.e., weather) or represent one-time-only fixed shifts in TFP over time (i.e., the variables reflecting institutional change). In this analysis, the coefficient of the spatial diversity variable is of most interest.

In addition to determining spatial diversity outcomes through their selection of varieties, farmers also choose the rate at which they adopt the new varieties embodying technological innovations. In this chapter, we use a measure of change in seed technology that is based on the rate of varietal turnover, VT[4], which is defined as:

$$VT_t = 1 \quad \text{for} \quad t = 1, \tag{10.3}$$

and $$VT_t = VT_{t-1} + S_k [V_{kt} = W_{kt} - W_{kt-1} \text{ if } W_{kt} - W_{kt-1} > 0,$$

otherwise $$V_{kt} = 0] \quad \text{for} \quad t > 1, \tag{10.4}$$

where V_k is the area share change for those varieties that have a positive sign and W_k is the area share of the kth variety in total wheat area sown (Jin et al. 2002). Equations (10.3) and (10.4) define seed technological change as the extent to which newly introduced varieties replace existing varieties. Assuming farmers are rational, variety replacement occurs if and only if the new variety has a higher "value" than the variety it replaces. A value improvement can be cost-reducing, yield-enhancing, or one that incorporates important consumption and/or market characteristics.

Since the farmer may simultaneously make production decisions that affect TFP, diversity, and technology adoption, an OLS regression of TFP on diversity and varietal turnover is likely to be problematic, due to correlation of the error term with these explanatory variables. To avoid the endogeneity of diversity and varietal turnover in the estimation of the TFP equation, we use an instrumental variables approach. Our strategy for identifying the effects of technology on TFP assumes that the technology delivered by the national and international research systems affects adoption (and hence both varietal turnover and diversity), but does not affect TFP except through the seeds (or basket of seeds) that farmers adopt. If these assumptions hold, we can use the following three variables as instruments for estimating variety turnover (VT); the investments made by the government in crop research (or more precisely the nation's stock of crop research); a measure of the germplasm flowing into each province from the research system and from international agricultural research centers; and yield potential, a variable representing the yield-increasing potential of technology generated by the research system (Equation 10.6).[5]

---

[4] *Varietal turnover is a measure of the time required for existing varieties to be replaced by new varieties in farmers' fields. As described in Chapter 2, it can also conceptually be viewed as diversity over time.*

[5] *Yield potential is defined to be non-decreasing. If a given major variety has the highest yield in a province in one year, yield potential in that province is calculated using that yield level. We assume that the yield potential will not fall, even in the case that farmers have stopped using the variety and all other varieties have lower certified yields in the following years.*

In production systems characterized by some amount of home consumption, as is the case for many of the regions in our study provinces, the consumption characteristics of the available varieties are often as important as the production traits, in the variety choices of farm households. Diversity outcomes can therefore be driven by both consumption and production considerations. We represent the production and technological aspects with some of the instrumental variables discussed above. We relate consumption aspects to the level of market activity, since those provinces with relatively less market involvement will have more households that consume their own output and thus that would also be concerned with the existence and quality of consumption characteristics. Accordingly, we utilize three additional instrument variables to enable identification for diversity: the level of co-integration of rice and maize markets, the ratio of market price to government quota price, and the difference between non-state purchases and purchases made through negotiated procurement (Equation 10.7).

We use a simultaneous, three-stage, least squares estimator to estimate the effect of diversity, technology, and other explanatory variables—including infrastructure, institutional change, environmental factors—on TFP. The empirical specifications of endogenously determined diversity and technology (*VT*), as well of the TFP model are as follows:

$$TFP_{ht} = f\,(VT_{ht}, Div_{ht}, Extension_t, Irrigation_{ht}, D_{90\text{-}95}, Weather\ Event\ Indices_{ht}, Environmental\ Quality, Provincial\ Dummies) + e1_{ht} \quad (10.5)$$

$$VT_{ht} = g_V\,(Extension_t, Irrigation_{ht}, D_{90\text{-}95}, Weather\ Event\ Indices_{ht}, Provincial\ Dummies, Research\ Stock_t, CG_{ht}, Yield\ Potential_{ht}) + e3_{ht}, \quad (10.6)$$

$$Diversity_{ht} = g_D\,(Extension_t, Irrigation_{ht}, D_{90\text{-}95}, Weather\ Event\ Indices_{ht}, Provincial\ Dummies, CG_{ht}, Yield\ Potential_{ht}, MktIntegration_{ht}, Priceratio_{ht}, Purchasediff_{ht}) + e2_{ht} \quad (10.7)$$

where *h* indexes provinces and *t* indexes time; total factor productivity (TFP), spatial diversity, and variety turnover (*VT*) are defined as above; *Extension* is a variable reflecting all expenditures made on the extension system and aggregated to the national level; *Irrigation* is measured as the ratio of irrigated land to cultivated land and accounts for changes in the availability of this input over time; $D_{90\text{-}95}$ is an indicator variable that equals 1 for the period between 1990 and 1995 and is included to measure the effect on TFP of factors unique to the period of market liberalization that China experienced in the early 1990s. We also include two environmental variables to account for production fluctuations due to the effect of flood and drought (*Flood Index* and *Drought Index*), jointly referred to as *Weather Event Indices* in Equations (10.5) to (10.7) but used separately in the analyses, as well as provincial dummies to control for unobserved fixed effects associated with each province.

Three variables in Equation (10.6), $Research\ Stock_t$, CG_{ht}, $Yield\ Potential_{ht}$, are used to identify the effect of technology on TFP, and five variables in Equation (10.7), CG_{ht}, $Yield\ Potential_{ht}$, $Mktintegration_{ht}$, $Priceratio_{ht}$, and

$Purchasediff_{ht}$, are used to identify the effect of diversity on TFP. Crop breeding research stock is used as a proxy for public investment in the creation of new varieties, since most research is either embodied in the seed itself or requires delivery by technology disseminating organizations. Here, given the study period, the primary organization is the extension system, the effect of which we account for separately. As described in Chapter 3 and following Pardey et al. (1992), the stock variable is calculated from research expenditures at the national level and a set of lag weights. The second measure, *yield potential,* is a variable representing the yield-increasing potential of technology generated by the research system and is defined as the maximum yield of any cultivated variety up to time period *t*. This variable should also explain the adoption of new seed technology but have no independent effect on TFP. Finally, we define a variable, *CG,* to represent the proportion of genetic material in China's germplasm for wheat contributed by the Consultative Group on International Agricultural Research (CGIAR) system, specifically the International Maize and Wheat Improvement Center (CIMMYT). This variable is created utilizing pedigree data for all major cultivated varieties in each period by assigning geometric weights to crossing material.[6]

Results

The econometric estimates of our model (Tables 10.1 and 10.2) generally perform well, and most of the parameter estimates are robust to specification changes. The system weighted R-squared statistics are greater than 0.92 for all the model specifications. Hausman (1983) tests for exclusion restrictions that examine the validity of instruments confirm that the choice of instruments used in the varietal turnover and diversity equations is statistically valid.[7] Statistically, our instruments have a high degree of explanatory power for technology and diversity, but do not affect TFP, except through their influence on technology in the form of varietal turnover or spatial diversity.

Determinants of Technology and Spatial Diversity

While the technology and spatial diversity equations are used mainly to provide consistent estimates of the endogenous variables in the TFP equation, they also provide interesting insights into the process of the technology and diversity creation in China. The positive and highly significant sign on the research stock variable in the technology equation indicates that investments in the research system have been effective (Tables 10.1 and 10.2). Moreover, a higher level of national research stock accelerates the pace of varietal turnover. If technology,

---

[6] *Specifically, 0.25/parent, 0.06/grandparent, and so on. CG contribution represents the proportion of germplasm with ancestors that are identified with source of origin from CIMMYT.*

[7] *A Lagrange multiplier test is used to test the exogeneity of the set of identifying instruments (Hausman 1983). Results indicate that the null hypotheses of no correlation between the exogenous instruments and the disturbance term from TFP equation cannot be rejected.*

Table 10.1 Results of TFP analysis using named variety-based diversity indices.

	Berger-Parker			*Margalef*			*Shannon*		
	TFP Equation	*VT Equation*	*BP Equation*	*TFP Equation*	*VT Equation*	*Marglef Equation*	*TFP Equation*	*VT Equation*	*Shannon Equation*
Shanxi	12.87	0.08	–4.24	–33.24	0.10	–0.34	–27.44	0.07	–0.20
	(0.38)	(0.43)	(2.26)**	(1.60)	(0.51)	(0.68)	(1.21)	(0.37)	(0.65)
Jiangsu	100.63	–0.40	–0.16	171.10	–0.40	–2.22	147.36	–0.40	–0.82
	(4.12)***	(2.14)**	(0.09)	(7.47)***	(2.14)**	(4.43)***	(6.99)***	(2.15)**	(2.70)***
Anhui	0.86	–0.02	1.01	36.21	–0.01	–0.60	35.19	–0.03	–0.28
	(0.09)	(0.25)	(1.39)	(4.32)***	(0.17)	(3.03)***	(4.10)***	(0.37)	(2.37)**
Shandong	44.93	–0.17	–0.87	57.94	–0.16	–0.76	53.57	–0.16	–0.34
	(3.81)***	(1.91)*	(1.01)	(5.64)***	(1.91)*	(3.33)***	(5.14)***	(1.90)*	(2.44)**
Henan	52.70	–0.22	–3.17	22.02	–0.21	–0.41	29.63	–0.23	–0.31
	(3.42)**	(–2.59)**	(4.01)***	(2.62)**	(2.48)**	(1.95)*	(3.12)***	(2.66)***	(2.39)**
Sichuan	53.64	–0.26	–4.79	42.57	–0.24	–1.64	64.89	–0.27	–1.13
	(2.38)**	(2.00)**	(4.26)***	(2.61)**	(1.86)*	(5.38)***	(3.19)***	(2.14)**	(6.14)***
Dtime	2.91	–0.11	0.06	4.58	–0.11	–0.04	6.28	–0.12	–0.02
	(0.26)	(–1.42)	(0.07)	(0.52)	(1.39)	(0.16)	(0.65)	(1.57)	(0.17)
Flood	–105.74	–0.06	–0.59	–103.23	–0.06	–0.29	–115.60	–0.06	0.03
	(4.52)	(–0.35)	(–0.35)	(–5.52)***	(0.34)	(0.65)	(5.78)***	(0.32)	(0.12)
Drought	–33.15	–0.21	0.18	–42.83	–0.21	0.24	–54.63	–0.21	0.35
	(1.40)	(–1.21)	(0.11)	(–2.25)**	(1.20)	(0.51)	(2.61)**	(1.22)	(1.25)
Irrigation	58.92	–0.42	–11.90	–215.97	–0.39	3.51	–146.31	–0.42	0.88
	(0.53)	(–0.57)	(1.64)	(3.00)***	(0.54)	(1.79)*	(1.92)*	(0.58)	(0.74)
Extension	0.068	0.007	–0.03	0.03	0.007	–0.02	0.11	0.007	–0.01
	(0.31)	(4.45)***	(1.26)	(0.18)	(4.48)***	(2.42)**	(0.60)	(4.38)***	(2.62)**

Table 10.1 Contd.

Table 10.1 Results of TFP analysis using named variety-based diversity indices.

	Berger-Parker			*Margalef*			*Shannon*		
	TFP Equation	*VT Equation*	*BP Equation*	*TFP Equation*	*VT Equation*	*Marglef Equation*	*TFP Equation*	*VT Equation*	*Shannon Equation*
VT	17.20			18.33			13.35		
	(2.61)**			(3.88)***			(2.30)**		
Research		0.013			0.013			0.013	
		(17.94)***			(17.8)***			(18.6)***	
CG		0.04	−0.23		0.04	−0.09		0.04	−0.03
		(1.48)	(0.95)		(1.52)	(1.26)		(1.57)	(0.64)
YieldPotential		0.003	−0.004		0.003	−0.002		0.03	−0.0004
		(2.49)**	(0.41)		(2.64)***	(0.63)		(2.38)***	(0.30)
Diversity	14.51			34.21			68.12		
	(3.49)***			(4.16)***			(4.40)***		
Mktint			−0.63			−0.54			−0.14
			(0.34)			(0.95)			(0.40)
Priceratio			−1.07			−0.63			−0.31
			(2.07)**			(3.85)***			(3.16)***
Purchasediff			0.13			−0.64			0.03
			(2.48)**			(3.85)***			(3.75)***

Note: ***, ** and * indicate statistical significance at 1%, 5% and 10%, respectively.

Table 10.2 Results of TFP analysis using morphological characteristics-based diversity indices.

	Berger-Parker			Margalef			Shannon		
	TFP Equation	*VT Equation*	*BP Equation*	*TFP Equation*	*VT Equation*	*Marglef Equation*	*TFP Equation*	*VT Equation*	*Shannon Equation*
Shanxi	–83.43	0.11	0.80	–86.70	0.11	0.15	–87.37	0.10	0.42
	(–2.57)**	(0.54)	(2.54)**	(3.47)***	(0.55)	(2.10)**	(3.09)***	(0.51)	(3.23)***
Jiangsu	162.82	–0.41	–1.30	200.20	–0.40	–0.40	207.05	–0.40	–1.05
	(4.09)***	(2.19)**	(4.27)***	(5.73)***	(2.17)**	(5.55)***	(4.36)***	(2.16)**	(8.17)***
Anhui	11.19	–0.01	0.13	56.70	–0.01	–0.17	49.21	–0.01	–0.34
	(1.05)	(0.18)	(1.06)	(3.72)***	(0.14)	(6.05)***	(2.98)***	(0.17)	(6.92)***
Shandong	69.70	–0.17	–0.70	85.21	–0.17	–0.19	98.05	–0.17	–0.57
	(3.06)***	(1.93)*	(4.88)***	(4.68)***	(1.92)*	(5.81)***	(3.45)***	(1.92)*	(–9.86)***
Henan	26.36	–0.21	–0.30	23.05	–0.21	–0.05	63.86	–0.21	–0.45
	(1.83)*	(2.45)**	(2.32)**	(2.30)**	(2.43)**	(1.67)*	(2.54)**	(2.49)**	(8.73)***
Sichuan	30.27	–0.24	–0.76	61.88	–0.23	–0.29	115.17	–0.24	–1.11
	(1.04)	(1.83)*	(4.16)***	(2.33)**	(1.80)*	(1.05)	(2.02)**	(1.87)*	(15.05)***
Dtime	9.40	–0.11	–0.13	6.67	–0.11	–0.03	11.87	–0.11	–0.10
	(0.60)	(1.37)	(0.94)	(0.62)	(1.35)	(1.05)	(0.90)	(1.36)	(1.73)*
Flood	–98.57	–0.06	–0.26	–123.99	–0.06	0.06	–115.40	–0.06	0.07
	(3.66)***	(0.34)	(0.93)	(5.63)***	(0.35)	(0.91)	(5.05)***	(0.35)	(0.62)
Drought	–34.65	–0.21	–0.09	–55.24	–0.21	0.07	–34.75	–0.21	–0.03
	(1.32)	(1.19)	(0.33)	(2.37)**	(1.19)	(1.06)	(1.50)	(1.19)	(0.27)
Irrigation	–308.57	–0.36	4.43	–319.98	–0.37	0.97	–341.77	–0.39	2.66
	(2.19)**	(0.49)	(3.75)***	(3.25)***	(0.50)	(3.41)***	(2.77)***	(0.53)	(5.28)***
Extension	–0.15	0.007	–0.004	–0.082	0.007	–0.001	–0.17	0.007	–0.002
	(0.57)	(4.49)***	(1.06)	(0.40)	(4.50)***	(1.07)	(0.71)	(4.49)***	(1.07)

Table 10.2 Contd.

Table 10.2 Results of TFP analysis using morphological characteristics-based diversity indices.

	Berger-Parker			*Margalef*			*Shannon*		
	TFP Equation	*VT Equation*	*BP Equation*	*TFP Equation*	*VT Equation*	*Marglef Equation*	*TFP Equation*	*VT Equation*	*Shannon Equation*
VT	26.26			28.52			30.36		
	(5.14)***			(6.21)***			(5.99)***		
Research		0.01			0.01			0.013	
		(17.70)***			(17.6)***			(17.8)***	
CG		0.05	–0.01		0.04	0.001		0.04	0.02
		(1.57)	(0.38)		(1.54)	(0.07)		(1.48)	(1.07)
YieldPotential		0.003	–0.001		0.003	–0.001		0.003	–0.002
		(2.61)**	(0.69)		(2.68)***	(2.36)**		(2.62)**	(3.11)***
Diversity	48.50			274.58			113.07		
	(1.58)			(3.03)***			(2.24)**		
Mktint			–0.94			–0.09			–0.42
			(2.49)**			(1.15)			(3.15)***
Priceratio			–0.23			–0.07			–0.12
			(3.12)***			(3.11)***			(3.40)***
Purchasediff			0.02			0.004			0.007
			(2.42)**			(1.89)*			(2.08)**

Note: ***, ** and * indicate statistical significance at 1%, 5% and 10%, respectively.

and the investment in R&D that lead to new advances in technology, are indeed key factors in the future growth of China's agriculture (Huang and Rozelle 1996), our results emphasize the necessity of maintaining the level and growth of investments in crop research and development.

The positive and significant signs of the yield potential variables in the varietal turnover equations (Tables 10.1 and 10.2, Row 15) suggest that farmers turn their varieties over more frequently when high-yielding wheat varieties are available in their provinces. This result also implies a farmer preference to adopt higher-yielding wheat varieties. Jin et al. (2002), however, show that this result does not hold in the case of rice or maize. The inconsistency could be due to a greater relative emphasis during the study period on non-yield traits or a relatively larger range in the availability of quality-related alternatives for rice and maize varieties.

Findings show that the difference between non-state purchases and negotiated procurement has a positive and significant impact on spatial diversity. The results are consistent for all definitions of spatial diversity and imply that a higher market demand leads to a greater diversification of wheat varieties. It is well known that, as a result of the government's procurement system, China's farmers in general would choose to sell relatively low quality grain to the government. Little incentives existed for the sale of quality and other characteristics, and wheat obtained through government purchases was consequently homogenous to a large degree. In contrast, non-state markets are likely to demand wheat with a much greater range of diversity due to special needs for various market and industrial uses. Accordingly, the demand for variety diversification can be expected to increase as the proportion of non-government market demand grows. Moreover, holding the time period constant, a province with a larger share of non-state market sales of wheat also tends to have more demand for varietal diversification. The unexpected negative and significant impact of the ratio of market price to government quota price on all diversity measures conflicts somewhat with a priori expectations.

Determinants of TFP

The coefficient estimates from the TFP specification are for the most part as expected and have low standard errors. Unfavorable drought and flood conditions consistently show a negative and largely significant effect on TFP. Similar to results obtained by Rosegrant and Evenson (1992), parameter estimates for irrigation are either insignificant or negative. Because the value of irrigation is embodied in the land input variable (areas with high land values have high levels of irrigation), its positive impact likely has already been taken into account. Moreover, as much of the newly irrigated area is on increasingly marginal land, the result could be only a very small overall, or even potentially negative, effect on wheat TFP, especially in the years immediately following installation. Since input decisions are often taken prior to negative weather events, however, impacts on TFP through input use may be limited.

With only one exception, all representations of spatial diversity used based on both taxonomies positively affect aggregate TFP. The pattern of results suggests that spatial diversity, whether measured in terms of named varieties or morphological characteristics, contributes to an increase in TFP. This result suggests strongly that support for the use of diverse materials in breeding research will have positive effects on future wheat productivity.

Another robust finding of our analysis is the large and positive influence of technology on TFP. This result holds over all model specifications. The positive and highly significant coefficient on the rate of varietal turnover shows that TFP increases as new technology is adopted by farmers. Both the positive contributions of China's research system and its success in pushing out yield potential imply that domestic investments in agricultural R&D have contributed to a healthy agricultural sector.

The role of extension appears to be less straightforward. The impact of extension can occur through its role in disseminating new seed technologies, as measured by the coefficient on the *Extension* variable in the technology and diversity equations, and through its provision of other services that enhance farmer productivity, as measured by the coefficient on the *Extension* variable in the TFP equation. The positive and significant coefficients on the extension variable in all of the varietal turnover equations demonstrate the importance of extension in facilitating farmer adoption (Tables 10.1 and 10.2). Surprisingly, however, extension appears to play little to no role in increasing (or decreasing) spatial diversity and plays no independent role in increasing the yield potential of adopted varieties. The latter result perhaps should not come as a surprise, given the reforms that have shifted extension from an advisory body to one that must be self-supporting, often through the sale of seed (Huang et al. 1999).

Conclusions

Our results establish a significant analytical link between aggregate productivity and spatial diversity. Interestingly, the significance of impact was stable over changes in both the specific representation of spatial diversity (richness, inverse dominance, evenness) and the taxonomy used (named variety, morphological grouping). The results reinforce previous findings that underline the importance of a continuing role of public investment in the agricultural research and extension system, to maintain and increase productivity. Our findings also highlight a specific avenue through which productivity gains can be improved. Spatial crop diversity in the mix of cultivated varieties has positively impacted wheat productivity. Attention to maintaining and increasing levels of diversity in the development of wheat varieties can provide a means of improving productivity. However, although the increasing focus on market forces and the growing importance of wheat quality and other characteristics may provide incentives for wheat scientists to expand sources of diversity in developing new crop varieties, other forces may be generating counterbalancing disincentives. Reliance on existing incentives to continue and expand the use of a diverse pool of breeding materials may thus be insufficient.

Additional studies on other crops and in other countries are warranted to verify both the methodology used and the conclusions drawn. Productivity impacts of crop diversity have been explored in previous studies; however, as Smale (2005) points out, due to mixed results, it is not clear how specific results are to locations and cropping systems. More research is required to clarify diversity-productivity relationships and the circumstances under which positive impacts take place.

REFERENCES

Ali, M. and D. Byerlee. 2002. Productivity growth and resource degradation in Pakistan's Punjab: a decomposition analysis. *Economic Development and Cultural Change* 50: 839-863.

Capalbo, S.M. and T.T. Vo. 1988. A Review of the Evidence on Agricultural Productivity and Aggregate Technology. In *Agricultural Productivity: Measurement and Explanation*. S.M. Capalbo and J.M. Antle (eds). Washington, D.C.: Resources for the Future.

Caves, D.W., L.R. Christensen, and W.E. Diewert. 1982. The economic theory of index numbers and the measurement of input, output, and productivity. *Econometrica* 50: 1393-1414.

Diewert, W.E. 1976. Exact and Superlative Index Numbers. *Journal of Econometrics* 4: 115-145.

Fan, S. 1997. Production and productivity growth in Chinese agriculture: New measurement and evidence. *Food Policy* 22: 213-228.

Hausman, J. 1983. Specification and estimation of simultaneous equations models. *Handbook of Economics,* Vol. 1. Z. Griliches and M. Intriligator (eds). Amsterdam: New Holland Publishing Company.

Huang, J.K. and S. Rozelle. 1995. Environmental stress and grain yields in China. *American Journal of Agricultural Economics* 77(4): 246-256.

Huang, J.K. and S. Rozelle. 1996. Technological change: Rediscovery of the engine of productivity growth in China's rural economy. *Journal of Development Economics* 49: 337-369.

Huang, J.K., R.F. Hu, F.B. Qiao, and S. Rozelle. 1999. Extension in Reform China, Working Paper, Center for Chinese Agricultural Policy, Beijing, China.

Jin, S., J. Huang, R. Hu, and S. Rozelle. 2002. The creation and spread of technology and total factor productivity in China's agriculture. *American Journal of Agricultural Economics* 84: 916-930.

Jin, S., C. Pray, J. Huang, and S. Rozelle. 1999. Commercializing agricultural research and fungible government investment: Lessons from China. Unpublished document.

Lin, J. 1992. Rural reform and agricultural growth in China. *American Economics Review* 82: 34-51.

McMillan, J., J. Whalley, and L. Zhu. 1989. The impact of China's economic reforms on agricultural productivity growth. *Journal of Political Economy* 97: 781-807.

MOA (Ministry of Agriculture). 1986-1997. *Statistical Compendium, China's Major Varieties.* Unpublished Statistical Volume. Beijing: Ministry of Agriculture.

Murgai, R. 2001. The Green Revolution and the productivity paradox: evidence from the Indian Punjab. *Agricultural Economics* 25: 199-209.

Pardey, P., R. Lindner, E. Abdurachman, S. Wood, S. Fan, W. Eveleens, B. Zhang, and J. Alston. 1992. The Economic Returns to Indonesian Rice and Soybean Research. Report prepared by the Agency for Agricultural Research and Development (AARD) and the International Service for National Agricultural Research (ISNAR).

Rosegrant M.W. and R.E. Evenson. 1992. Agricultural productivity and sources of growth in South Asia. *American Journal of Agricultural Economics* 74: 757-761.

Rozelle, S., C. Pray, and J. Huang. 1997. Agricultural research policy in China: Testing the limits of commercialization-led reform. *Contemporary Economic Policy* 39: 37-71.

Smale, M. 2005. Chapter II: Economics literature about crop biodiversity: Findings, methods and limitations. In *The State of Economics Research: Sustainable Management of Crop and Livestock Biodiversity*. Draft prepared for the System-wide Genetics Resources Program.[??]

SPB (State Price Bureau). 1988-1992. Quanguo Nongchanpin Chengben Shouyi Ziliao Huibian (National Agricultural Production Cost and Revenue Information Summary). Beijing: China Price Bureau Press.

Stone, B. and S. Rozelle. 1995. Foodcrop Production Variability in China, 1931-1985. *The School for Oriental and African Studies, Research and Notes Monograph Series*, Volume 9, London, August 1995.

Tang, A.M. and B. Stone. 1980. *Food Production in the People's Republic of China.* International Food Policy Research Institute (IFPRI) Report.

Wiens, T.J. 1982. Technological Change. *The Chinese Agricultural Economy*. R. Barker and R. Sinha (eds.). Boulder, Colorado: Westview Press.

Wen, G.Z. 1993. Total factor productivity change in China's farming sector: 1952-1989, *Economic Development and Cultural Change* 42: 1-41.

Chapter 11

Cost Function Analysis of Wheat Diversity in China

E. Meng, M. Smale, S. Rozelle, R. Hu, and J. Huang[1]

ABSTRACT

A cost function approach is used to assess the impact of wheat diversity on the allocative efficiency of wheat production in China. Using this approach enables an examination of the marginal economic cost or benefit of changes in wheat diversity, as well as the identification of its impacts on input allocation. Measures of richness, inverse dominance and evenness are tested for morphological characteristics of wheat varieties in seven key wheat-producing provinces in China from 1982 to 1995. The results of the analysis show that while increased evenness across morphological characteristics increases total costs per hectare, spatial diversity can effectively substitute for specific inputs such as pesticides and labor. Factors determining spatial diversity in named varieties are also identified to provide a clearer understanding of the implications of changing policies relating to those inputs.

Research to examine the effects of crop diversity on productivity initially utilized a primal approach to estimate the marginal effects of diversity on technical efficiency (Smale et al. 1998; Widawsky and Rozelle 1998). A primal approach, however, does not allow an explicit consideration of allocative efficiency. In an analysis of various dimensions of the diversity of modern spring bread wheat varieties cultivated in developing countries over a 30-year period since 1965, Smale et al. (2003) found significant improvement in performance over time and across different environments of input-use efficiency, genetic resistance to disease, and heat and drought tolerance. Although the findings suggest a relationship between changes in diversity and changes in any productivity-related performance indicator, the analysis does not attempt to estimate such a relationship.

[1]*This chapter draws heavily on Meng et al. (2003).*

In this chapter, we link indices of spatial diversity based on groupings statistically classified using morphological characteristics for major wheat varieties from seven major wheat producing provinces in China.[2] We then link these diversity indices to economic decisions through the estimation of a cost function for wheat. By using this approach, we are able to examine the marginal economic cost (or benefit) of wheat diversity, as well as its effects on input allocations. The next section presents a methodology and data for the estimation of the cost function, followed by a brief description of wheat diversity indices used in the analysis. Estimation results are then presented, and we conclude the chapter with a brief discussion of policy implications.

Wheat Diversity in China

As discussed in Chapter 2, crop populations can be classified in many different ways, including the use of variety names, morphological characteristics, or other criteria that farmers use to describe them. They can also be classified using genealogies as recorded by plant breeders or by the genetic identity that molecular analysis reveals. In this chapter, we utilize as the basis for our diversity indices groups derived statistically from variation in plant morphological characteristics (Franco et al. 1998).

Indices of richness, inverse dominance and evenness based on morphological characteristics for each of the seven major wheat-producing provinces of China have been individually examined in Chapter 6. The diversity indices constructed from data on named wheat varieties consistently attribute a higher level of diversity to the provinces of Shanxi, Anhui, and Hebei than to the other provinces included in the analysis. Both the richness and dominance indices for named varieties indicate that Sichuan Province is the least diverse of all the provinces included in the analysis. Indices based on morphological groups concur on the relatively low diversity levels of Sichuan Province, although there are differences in the relative order of the remaining provinces. The evenness indices reinforce the findings of the richness and abundance indices.

Impact of Diversity on Costs of Production

The structure of agricultural production and productivity growth in Chinese agriculture has been described in detail in Chapter 10. Largely for data-related reasons, most previous studies examining productivity growth and efficiency in Chinese agriculture have approached the analysis from the primal side. Two studies have taken a dual approach to examine production efficiency. As a result of market distortions caused by sociopolitical and institutional

[2] *A major variety in our sample is any variety that has a cultivated area of at least 100,000 mu (6,667 hectares) in a province. Although our database does not have full coverage of all varieties in each province, the proportion of area covered by "major" varieties exceeds 90% in each province.*

constraints, Wang et al. (1996) questioned the appropriateness of assuming profit-maximizing behavior in China and estimated a shadow frontier profit function with household data to separate the effects of technical and allocative efficiency. Fan (1999) accepted the assumption of cost minimization, but contended that rice farmers in Jiangsu allocated inputs (fertilizers, pesticides, labor, and machinery) according to their shadow prices.

Following the methodology of Fan (1999), we assume that wheat farmers minimize costs, and we utilize his methodology and results to estimate a shadow cost function for wheat that takes into consideration distortions in Chinese input markets. Using the Shannon index of evenness based on morphological groups of wheat varieties as our index of crop diversity, we test the hypothesis that the spatial diversity of wheat affects the total cost of wheat production and input allocations in seven major wheat-producing provinces of China. Using this framework, we estimate the marginal cost of promoting a more "equitable" distribution of wheat varieties in terms of the economic efficiency of wheat production.

Due to the complex nature of the relationship between wheat diversity and the costs of production and input cost shares, as with the relationship between wheat diversity and total factor productivity (TFP) in Chapter 10, the overall effect on per unit costs of production *a priori* is ambiguous. However, expected effects of diversity on the cost shares of specific inputs for wheat production can be hypothesized. Taking into consideration the beneficial impact of wheat diversity in a given area on the spread of disease and pest damage, we expect *a priori* for our diversity measure to have a negative effect on the cost share of pesticides. With respect to the cost share of labor, a greater amount of variation in the wheat varieties cultivated could increase the efficiency of available labor and reduce labor costs by allowing households the flexibility to allocate their labor more efficiently during busy periods, particularly during planting and harvesting. On the other hand, it could also be argued that a uniform wheat crop facilitates more efficient labor use by the farmers, since cultivating a more varied wheat crop may well create more work.

The effect of spatial diversity on machinery expenditures is expected to be positive, if increased morphological variation of cultivated wheat negatively impacts any scale efficiencies associated with machinery use. If morphological variation mitigates time-related bottlenecks associated with machinery access during busy times, a negative effect is also possible. Smale et al. (2002) provide evidence from numerous studies confirming the input efficiency and responsiveness of modern spring wheat cultivars, particularly with respect to nitrogen fertilizer. Breeding under medium to high nitrogen conditions has resulted in improved yield performance of those varieties, whether grown under high nitrogen or low nitrogen conditions. Although the pool of varieties cultivated in China was not included as part of the Smale et al. (2002) study, many breeding objectives and methods are not dissimilar, and we assume that the *a priori* relationship between spatial diversity and fertilizer cost shares will be negative.

Model Specification

Since spatial diversity indices are constructed from data on area shares planted to varieties, their potential endogeneity relative to input use decisions must be addressed. We do so with an instrumental variable approach. Furthermore, since provinces rather than farmers are the chosen unit of analysis, and farmers are unable to observe the effects of crop diversity on risk, a risk-neutral decision-making framework is used. We specify our model as Cobb-Douglas with the following general form (suppressing subscripts for year and province):

$$C = C(y^{\circ}, r \mid \mathbf{v}, \delta, \mathbf{z}, t) \qquad (11.1)$$

$$s_i = s_i(\delta, \mathbf{v}, \mathbf{z}, t) \qquad (11.2)$$

$$\delta = \delta(\delta_m, \delta_b, \mathbf{v}, R, \mathbf{z}, \mathbf{d}) \qquad (11.3)$$

We hypothesize that total wheat costs per hectare (C) are determined by the predicted level of wheat output (y°) and a vector of shadow input prices (r), conditioned on genetic diversity (δ), the experimental yield potential of cultivated wheat varieties and age of wheat cultivars grown (**v**), and a vector of policy and environmental variables (**z**). The latter vector includes shifter variables for policy regimes from 1982-84 and 1991-95 representing, respectively, the early reform period and a period of market liberalization. The vector also includes a multiple cropping index reflecting the number of crops cultivated per year as well as variables measuring the amount of arable land affected by erosion, drought, and flooding. Finally, a separate variable for the overall stock of government investment in agricultural research is included. This variable does not vary across provinces and reflects combined expenditures on extension and breeding research, as well as other management and administrative costs.

Shares for i inputs (s_i), labor, fertilizer, pesticide, and machinery, are specified as constant in the Cobb-Douglas cost function, but are also conditioned on the shifter variables specified in the cost function. A time trend and squared time trend are included in cost and share equations, to estimate neutral technological change.

The diversity equation (11.3) expresses the evenness (Chapter 2) in the spatial diversity of wheat morphological groups in China, as driven in part by the same environmental factors that affect the cost of wheat production (**z**) and the weighted average yield potential and weighted average age of cultivars accounting for 80% of the sown wheat area (**v**). Instruments in the equation are richness and inverse dominance of diversity, represented by the Margalef and Berger Parker indices calculated from area shares of named cultivars (δ_m, δ_b). Other variables included in the equation are the total level of research expenditures (R) and provincial fixed effects (**d**). This equation states that, environmental factors and provincial effects held constant, the morphological evenness of a wheat crop in any given year and province is determined by the availability of germplasm in the wheat research system, the richness and relative abundance of named cultivars grown by farmers in that year, and parameters related to the past diffusion of cultivars. Farmers individually choose to allocate their land among the pool of wheat varieties that are available to them, but in the aggregate, their choices determine the distribution over space of morphological traits. We hypothesize that the evenness of this distribution in terms of morphological traits, in turn, affects wheat productivity through wheat costs of production and input allocations.

Data

We used panel data on input and output prices, expenditures, environmental conditions, and government investments for the provinces of Anhui, Hebei, Henan, Jiangsu, Shanxi, Shandong, and Sichuan from 1982 to 1995, much of which is common to the data described and used in Chapter 10. Much of the data for quantities and total expenditures of both major and miscellaneous inputs comes from a national State Price Bureau (SPB) data set collected over a 20-year period from a representative sample of more than 20,000 households and from provincial surveys (SPB). Original data from surveys in China also contribute to the cost data used in this study (Jin et al. 2002). Additional data on wheat variety cultivation and production are calculated from publications of the Ministry of Agriculture and China's statistical and agricultural yearbooks (MOA; NSBC).

Estimation Results

We estimate the simultaneous system specified using LIMDEP's three-stage least squares program in an approach that largely follows that used by Antle and Pingali (1994) to analyze the effects of pesticide use on health and the cost of rice production in the Philippines. Restrictions on input prices within the cost function as well as cross-equation restrictions are imposed, and the cost share equation for machinery was dropped. The variable for wheat output used in the system is predicted in a single-equation ordinary least squares regression with sown area, lagged output price, a variable representing irrigation infrastructure, variety-specific (**v**) factors, and the vector of policy and environmental variables (**z**).

The estimated coefficients of the system are presented in Table 11.1. Input prices and output, the conventional variables in the cost system, are significant and of the sign consistent with economic theory. As expected, environmental degradation resulting from erosion, flooding, and drought increases costs of production. The area-weighted experimental yield potential of varieties cultivated has a similar effect, perhaps due to farmers acting on the perception that these varieties require a higher level of purchased inputs. This variable is also positive and significant in the fertilizer cost share equation, a finding that, while not expected *a priori*, can be interpreted as reinforcing this conclusion.

Table 11.1 Results of 3SLS estimation of cost and share equations with endogenous genetic diversity.

Explanatory variable	*Cost*	*Labor*	*Fertilizer*	*Pesticide*	*Genetic diversity*
Constant	7.279**	0.369**	0.432**	0.112	–41.2**
Time	–0.0836**	0.0131*	–0.0334**	0.00486**	–
$Time^2$	0.00449**	–0.000423	0.000360	–0.000163**	–
Erosion	0.0418**	0.0547**	–0.0244	–0.00298**	7.975**
Flood	0.169**	–0.0108	–0.00690	0.00426	1.171
Drought	0.180**	–0.0243	0.0189	–0.0128	0.189
Multiple cropping index	–0.311**	0.165**	–0.00189	0.00604**	–4.92**
Variety yield potential	0.00189**	–0.000892**	0.000635**	–0.000246**	0.0306**
Variety age	–0.00854	0.00265	0.0205**	–0.00483**	1.542**
Policy regime 1	–0.0333	0.0291*	–0.0446**	0.00469	–
Policy regime 3	–0.0593*	–0.0274*	0.0000192	0.00294	–
Wheat output	0.0000938**	–	–	–	–
Wage	0.369**	–	–	–	–
Fertilizer price	0.432**	–	–	–	–
Pesticide price	0.112**	–	–	–	–
Machinery price	0.0874**	–	–	–	–
Variety evenness (morphological groups)	0.000167**	–0.000101**	0.0000932**	–0.000021**	–
Variety richness (named varieties)	–	–	–	–	1.07**
Inverse variety dominance (named varieties)	–	–	–	–	–0.063
Research investment	–	–	–	–	–0.738**
Anhui	–	–	–	–	22.3**
Hebei	–	–	–	–	18.47**
Henan	–	–	–	–	21.1**
Jiangsu	–	–	–	–	28.6**
Shandong	–	–	–	–	22.6**
Shanxi	–	–	–	–	3.17**
n = 98	–	–	–	–	–
F-significance	0.00	0.00	0.00	0.02	0.00
Wald	0.00	0.00	0.00	0.00	0.00

* significant at 0.10 level with Z test

** significant at 0.05 level with Z test.

Estimation results show a positive and significant effect of evenness across morphological characteristics on total per-hectare wheat production costs. Evenness in diversity does, however, have a negative and significant effect on the cost shares for labor and pesticides. Morphological evenness (evenness among morphological groups) thus appears to contribute to higher per-hectare costs of production, but cost savings for particular inputs. The latter result suggests that a more equitable distribution of morphological traits provides a natural means of defense against pests and thus reduces the expenditure share of pesticides. Spatial diversity would, in effect, function as a substitute to chemically-based or other purchased methods of biotic control and enable a reduction in both the level of use and the pesticide expenditures required.

With respect to the cost share of labor, the results indicate that increasing the equity of distribution in morphological traits alleviates labor bottlenecks and inefficiencies during key periods of wheat planting and harvesting. The effects on cost shares of labor would be less relevant or measurable without a diversity index based on characteristics that reflect labor use, such as crop maturity. Maturity, one of the morphological traits utilized for the determination of variety groups, is an important consideration for farmers in multiple-cropping systems.

Findings from the diversity equation suggest a positive relationship between richness in named varieties and evenness in groups categorized by morphological traits. The presence of relatively unproductive land, particularly land prone to erosion, increases diversity, possibly due to a shortage in the supply of varieties specifically developed and targeted for those types of growing conditions relative to those focused on more optimal growing conditions. Evenness in morphological diversity is also significantly increased by a slower turnover in cultivated varieties.

Research expenditures are found to have a negative effect on diversity, indicating that formal breeding programs regularly draw from a pool of morphological traits that are similar. Results also show a negative influence of the multiple cropping index variable, suggesting the existence of a limited pool of varieties with the required characteristics suitable for a specific cropping rotation. It is also consistent with the findings for Australia in Chapter 7 that suggested a wider range of cropping options for farmers likely results in less attention or less need for attention to the diversity within their wheat crop. Finally, indicator variables for the provinces also show that, controlling for policy and environmental variables, all provinces still exhibit a higher level of diversity than Sichuan.

The effect of factors other than diversity on individual cost share estimations is also of interest. A more intensive level of cropping increases cost shares for labor and pesticide, but is not significant for fertilizer. This result is perhaps not surprising, given that fertilizer is not equally applied to all crops in a rotation. Environmental variables are not significant, with the exception of an increase in the cost share of labor and a decrease in the cost share of pesticide for erosion-prone land. Land that is likely to erode, such as small, mountainous plots, requires more labor time both to access and to cultivate. The negative effect on the cost share of pesticide suggests that the presence of relatively more diversity in these areas may decrease the need for and/or use of chemical pesticides. It is also likely that a lower level of wheat productivity in these environments may make the use of purchased pesticides uneconomical for farmers. Experimental yield potential has differing effects on input costs and may reflect the use of genetic material that broadens resistance and eases labor requirements, but that is perceived to require more regular and intensive fertilizer use.

Policy Implications

In response to the emphasis placed on food security and grain production by the Chinese central government, one of the top priorities in Chinese agricultural research is the development of new varieties that continue to push out the yield frontier. Increases in yield potential can be achieved through the general influx of new genetic materials. Yield gains can also be obtained by developing varieties with characteristics that are adaptable to less than optimal environmental conditions. The targeted inclusion of genetic materials that reinforce or replace sources of resistance to existing and new diseases and pests is also an effective strategy in protecting yield gains that have been achieved (Marasas et al. 2004) and in providing yield stability (Gollin 2006). Diversity thus plays an indirect role in more technically efficient wheat production by advancing scientific gains in breeding, particularly those targeted for specific diseases and environments.

Yield potential remains distinct, however, from on-farm crop productivity. The spatial distribution of genotypes across a crop-producing area affects disease resistance and tolerance to abiotic stresses, which in turn affects the average yields realized in the region. Since the spatial diversity of the wheat crop is determined not only by the genetic diversity incorporated through breeding but the variety choice decisions of farmers, we have modeled the productivity-diversity relationship as endogenous. While this may represent both a theoretical and empirical advance in our conceptual understanding of these issues, it remains difficult with this specification to disengage the effect of breeding from the effect of variety choice. Both factors are represented here in one spatial diversity index.

Although econometric results indicate that evenness in morphological groups is a positive factor in overall costs per hectare of wheat production, the relationship of morphologically represented diversity to specific input use carries potentially important cost-saving implications. If the introduction of new sources for pest and disease resistance has simultaneously resulted in increased levels of measured diversity, interaction with other required production inputs may have also changed. Diversity may thus contribute to a more efficient use of inputs, such as pesticides, which otherwise would have been required for a similar level of production stability.

REFERENCES

Antle, J. and P. Pingali. 1994. Pesticides, productivity and farmer health: A Philippine case study. *American Journal of Agricultural Economics* 76: 418-430.

Fan, S. 1999. Technological change, technical and allocative efficiency in Chinese agriculture: The case of rice production in Jiangsu. EPTD Discussion Paper No. 39, Washington: International Food Policy Research Institute.

Franco, J., J. Crossa, J. Villaseñor, S. Taba, and S.A. Eberhart. 1998. Classifying genetic resources by categorical and continuous variables. *Crop Science* 38: 1688-1696.

Gollin, D. 2006. *Impacts of International Research on Intertemporal Yield Stability in Wheat and Maize: An Economic Assessment.* Mexico, D.F.: CIMMYT.

Jin, S., S. Rozelle, R. Hu, and J. Huang. 2002. The creation and spread of technology and total factor productivity in China's agriculture. *American Journal of Agricultural Economics* 84(4): 916-930.

Marasas, C.N., M. Smale, and R.P. Singh. 2004. *The Economic Impact in Developing Countries of Leaf Rust Resistance Breeding in CIMMYT-related Spring Bread Wheat.* Economics Program Paper 04-01. Mexico, D.F.: CIMMYT.

Meng, E., M. Smale, R. Hu, S. Rozelle, and J. Huang. 2003. Wheat genetic diversity in China: measurement and cost. In *Agricultural Trade and Policy in China: Issues, Analysis and Implications.* Rozelle, S. and D.A. Sumner (eds.) Ashgate Publishing Ltd.

MOA (Ministry of Agriculture). 1986-1997. *Statistical Compendium, China's Major Varieties.* Unpublished Statistical Volume. Beijing, China: Ministry of Agriculture Press.

NSBC (National Statistical Bureau of China). 1980-2002 (various issues). *Zhongguo Tongji Nianjian [Statistical Yearbook of China].* Beijing, China: China Statistical Press.

Smale, M., J. Hartell, P.W. Heisey, and B. Senauer. 1998. The contribution of genetic resources and diversity to wheat production in the Punjab of Pakistan. *American Journal of Agricultural Economics* 80: 482-493.

Smale, M., M.P. Reynolds, M. Warburton, B. Skovmand, R. Trethowan, R.P. Singh, I. Ortiz-Monasterio, and J. Crossa. 2002. Dimensions of diversity in modern spring bread wheat in developing countries since 1965. *Crop Science* 42: 1766-1779.

Smale, M., E. Meng, J.P. Brennan, and R. Hu. 2003. Determinants of Spatial Diversity in Modern Wheat: Examples from Australia and China. *Agricultural Economics* 28(1):13-26.

SPB (State Price Bureau). 1988-1992. Quanguo Nongchanpin Chengben Shouyi Ziliao Huibian (National Agricultural Production Cost and Revenue Information Summary). Beijing: China Price Bureau Press.

Wang, J., E. Wailes and G. Cramer. 1996. A shadow-price frontier measurement of profit efficiency in Chinese agriculture. *American Journal of Agricultural Economics* 78: 146-156.

Widawsky, D. and S. Rozelle. 1998. Varietal diversity and yield variability in Chinese rice production. In *Farmers, Gene Banks and Crop Breeding: Economic Analyses of Diversity in Wheat, Maize, and Rice,* M. Smale (ed.). Norwell, Massachusetts: Kluwer Academic Publishers.

Chapter 12

Overview and Policy Implications

E. Meng and J.P. Brennan

To assess and synthesize the implications of the findings of the analyses presented in previous chapters, we present here an overview of the key findings regarding the determination of varietal diversity and the impacts of diversity changes on wheat productivity. Key lessons from the comparison of wheat production systems in China and Australia are also drawn out. The progress made in advancing the conceptual framework for the economic analysis of diversity proposed in the initial chapters is then addressed. We conclude with a discussion of policy implications.

OVERVIEW

Determinants of Varietal Diversity

Our ability to understand the factors that influence diversity outcomes is crucial to advancing the economics of crop diversity. Several chapters in this volume explore or test hypotheses on the determinants of diversity outcomes at varying levels of analysis. Chapters 3 and 4 also reinforce the importance of the policy environment in influencing decisions that affect crop diversity. The national policy environment—whether in China, with its continuously shifting balance between central and local governments and the introduction of individual incentives, or in Australia, with its national wheat marketing policy—has a significant effect on the incentives for and implications of wheat diversity. In addition, policies related to the structure of the seed industry, the role of new production technologies, output marketing structure and pricing arrangements, and other politico-bureaucratic institutions have direct and indirect effects on varietal diversity in a particular region or state. Results in Chapter 8 show that the deregulation of the Australian wheat market has been associated with a reduction in richness of the variety mix, but an increase in the level of spatial diversity by other measures. From the Chinese data, we see that the effect of market liberalization on the richness of the wheat varieties grown by farmers is positive, while the period of the household reform system is associated with greater dominance of the leading varieties and more uneven distribution of variety area shares.

The attitude of wheat breeders to diversity and their approach for handling it in their programs have a direct effect on varietal diversity (Chapter 5). Breeders play a key role in the supply of genetic diversity to farmers and affect the demand for genetic diversity from gene banks. While breeders in Australia reported that the genetic diversity of the materials they worked with influenced activities at all stages in their programs, they expressed concern that the current funding arrangements and industry pressures were forcing greater emphasis on the short-term goal of releasing new varieties. Such pressures are almost certain to affect the genetic diversity of the flow of new improved varieties into the future.

In Chapter 8, the hypothesis that variation in spatial diversity can be explained by the economic factors related to the supply and demand for varieties and the physical features of the production environment is tested. The importance of variety characteristics such as yield, quality, and maturity in explaining the diversity of wheat varieties grown by farmers is evident in both the Chinese and Australian systems. However, while data suggest an apparent trade-off between yield potential and the richness of wheat varieties grown in China, higher yield potential is consistent with greater spatial diversity in New South Wales (NSW).

In both systems, a slower rate of variety turnover in the field is positively related to spatial diversity in wheat; where varieties are changed more rapidly, there is greater uniformity at any given time, as farmers rapidly replace older varieties with newer ones. Since variety turnover is a principal defense mechanism against the overcoming of genetic resistance to pathogens in systems of modern varieties, this result suggests the potential for a yield trade-off associated with increased spatial diversity over the longer term, unless breeding priorities to include resistance genes in all released material are in place. In NSW, a more rapid rate of variety release and a higher proportion of locally bred material enhance spatial diversity.

Both sets of analyses in Chapter 8 confirm the major role of the physical environment in determining spatial diversity. Moisture regimes and the probability of having to sow late influence the number of varieties grown per unit of area in NSW. Evenness in the distribution of soil types is related to the evenness of area allocation among wheat varieties, as well as to the dominance of the leading wheat variety. Erosion, salinity, irrigation and cropping intensity are key factors in the Chinese setting. Shire and province fixed effects are significant determinants in both NSW and China. Though modern wheat varieties are less closely adapted to environments and specific settings than landraces, which evolve through farmer and natural selection within the local environment, the heterogeneity of the production environment influences the performance of the genetic materials that the breeding programs and the seed system provide.

The level of geographical aggregation at which diversity is measured is an important factor in determining the actual extent of diversity. The increase in measured diversity, as the range of environments included in a region broadens, is evident in the Australian data (Chapter 7). These findings highlight the importance of ensuring similar levels of aggregation, when making comparisons of diversity. The limited availability of data based on agro-ecological boundaries as opposed to political units also adds complexity to the analysis. As policies are also largely based on political units, not production environment units, straightforward policy analysis, recommendations, and actions are not easily forthcoming.

As confirmed in earlier on-farm research, farmers' choices of crop variety reflect their demand for the variety traits that confer economic value to them. In a predominantly commercial system like Australia's, there is less variation related to farm management practices and production objectives than in systems characterized by a higher degree of subsistence. In commercial systems, traits such as days to maturity, expected yield, grain quality, and resistance to lodging are essentially fixed from the viewpoint of the farmer, though they are malleable to change over time through plant breeding. The primary utilization of wheat output is for market sales, and factors such as yield differences and price premiums for grain quality determine relative profitability of different varieties. In a semi-commercial system such as that in China, yield differences are important in determining the relative value of varieties to households, but relative value also depends on household characteristics that affect their on-farm wheat consumption and access to markets (Chapter 9). Despite its setting in a system dominated by modern varieties, the economic framework for modelling farmer decision-making in China more closely parallels farmers cultivating traditional varieties in centers of diversity than does the framework for Australia. This is due to the role of home consumption, the existence of imperfect markets and information, and the multiple household objectives that a wheat variety or set of wheat varieties must satisfy.

Farmers' choices in any production system are constrained by the supply of seed for those varieties. New wheat germplasm is supplied to farmers in a commercial system as the product of public and private breeding programs, rather than their own on-farm selection practices. The supply of varieties is determined by a complex of factors, including past investments in research, the flow of germplasm and varieties from other programs, and policies affecting variety release and seed sales and distribution.

Impacts of Diversity Changes on Productivity

The research presented in Chapters 10 and 11 contributes to the body of literature on the economic value of crop diversity, by attempting to quantify the relationship between crop diversity and productivity. It builds on existing research examining returns to research from yield gains in new agricultural technologies. The rationale and approach of the productivity studies also broadly parallel research on the economic value of landraces in the sense that, as an initial estimate, a lower bound on the total value can be approximated by the direct use value of the resources to users. Users in the case of landraces are farmers in centers of crop domestication and diversity, and both revealed and stated preference methods can be utilized to elicit their valuation of the genetic resources. At a more aggregate level, diversity in the spatial distribution of wheat varieties also has economic value through its effects on crop productivity. Users are those benefiting from the productivity gains made possible by diverse crop genetic resources, and an estimate of contributions to productivity can provide a starting point for a better idea of the value of the resources.

The analyses of both the total factor productivity of wheat and cost of wheat production reveal that diversity, as measured by an index of richness and relative abundance in morphological characteristics, is a significant factor, albeit a relatively second-level one, in determining productivity growth. In the case of total factor productivity, diversity contributes directly, controlling for other hypothesized factors, but in the case of costs of production, the contributions appear indirectly through decreased cost shares of labor and pesticide use. Through its impact on productivity gains and decreased costs of production, the economic value of the contribution of spatial diversity to productivity begins to take a slightly more concrete form.

It is not immediately clear why the direct effect of diversity on costs of production does not better reflect the observed effect on total factor productivity, since the analyses should mirror each other according to economic theory. The regulated semi-commercial marketing system may contribute to obscuring the expected relationship between the two approaches. The use of a Tornquist-Theil index of TFP in the primal model rather than an explicit functional form (although under certain assumptions the Tornquist-Theil index corresponds to a translog), compared to a Cobb-Douglas functional form in the dual may also partially explain why the estimation outcomes in Chapter 10 and Chapter 11 do not lead unambiguously to the same conclusions. Furthermore, differences in the specifications of each model and in the chosen instruments likely play a role in explaining the differences in model outcomes between the cost function and TFP approaches.

Empirical results from both the TFP and cost function models, however, confirm the significant role of specific government policies in shaping diversity outcomes. These include not only policies directly related to priorities in agricultural research, but also those affecting infrastructure development and prices. The results strongly suggest that policies favorable to the maintenance and increase of crop diversity levels are also likely to positively influence the productivity of the wheat industry in China.

Comparison between Australia and China

The analyses presented in the chapters of this volume provide some interesting comparisons between Australia, with its highly commercialized level of wheat production, and China, with its mix between subsistence and commercial systems. Specifically, for wheat diversity at the national level, Australia is characterized by a relatively centralized decision-making structure for such issues such as the development and release of varieties and wheat quality. Standardized grades and commercial requirements are utilized as policy instruments to achieve predetermined national targets for commercial sales and export market orientation. In contrast, although the focus on wheat quality has increased and quality traits have taken on greater importance as breeding objectives, China's wheat industry is not as uniformly export-driven nor shaped by market forces. Decision-making on many similar issues at the national level is less centralized in China and has less influence on farm level variety decisions, as compared with the range of price and market-oriented incentives available in Australia. Institutions and factors affecting policy implementation at the provincial and county levels appear to carry more significance for variety supply and diversity outcomes, and the direct influence of wheat consumption characteristics on production decisions at the household level continues to contribute to variation in diversity outcomes.

Although biotic stresses show considerable heterogeneity in both countries, growing environments in Australia are less heterogeneous than those in China. There is a single, common growing season in Australia, compared with the winter, spring, and facultative wheat systems followed in China, with spring habit wheat being planted in either spring or fall, depending on locality. Similarly, the range of end users and uses is not as varied in Australia. One implication is that Australia may have a greater need to ensure adequate diversity levels through coordinated breeding objectives than China, where other forces favor diversity.

The above comparisons apply at the national level. At more disaggregated levels within each state or province, we observed fewer differences in factors affecting diversity, given that the major differences occur between provinces in China, rather than within provinces. Thus, at the state or province levels, and in particular at even more disaggregated local district or shire levels, there is a potential for localized problems and thus a greater need for awareness in both countries of the importance of maintaining or enhancing crop genetic diversity. Targeted policies, or awareness of the potential impacts of existing policies at the very least, may also be needed at disaggregated levels to ensure desired diversity outcomes. The pattern of higher diversity across geographical units compared to levels within units also raises the question of what potentially conflicting determinants are in play within and across the geographical units and where the most appropriate entry points for policy interventions should be.

ISSUES AND IMPLICATIONS

Progress in Advancing the Conceptual Framework for Economic Analysis of Diversity

The research in this volume includes the outline of a broad framework that incorporates many dimensions of economic analysis of crop diversity and contributes to this framework with additional applications of diversity analysis focused on modern wheat systems. These examples of the application of measurement tools for crop diversity across space and time contribute to the stock of information leading to both a better understanding of the concepts involved and a more detailed exploration of relationships among measurement methods. The results of this research also contribute to an improved understanding of the factors influencing spatial diversity in wheat by examining diversity outcomes and determinants at a range of different levels of analysis.

As we have seen, the data requirements to analyze different aspects of diversity vary widely, and in many situations it is the availability of data that represents the key constraint to, or indeed is the determining factor for, the selection of the diversity concept and measure. For example, one representation of latent diversity requires data on variety pedigrees as well as area shares, while for other measures only data on area shares are needed. However, some of the measures used here can be and have been applied in diverse scenarios, so that comparative assessments can be made regarding observed differences in levels of diversity.

Spatial versus Temporal Diversity

Temporal diversity can substitute to some extent for spatial diversity, for example as a means of reducing crop losses due to the evolution of pathogens, but we have largely focused on the application of spatial indices for analyzing the distribution and significance of diversity at specific points in time. Temporal indices, particularly a temporal index enabling a dynamic analysis of productivity—a category for which there now seem to be very limited, if any, measures available—will generate substantive insights of a different nature.

Spatial diversity and temporal diversity are clearly not perfect substitutes. An extreme manifestation of temporal diversity would involve the annual turnover of all varieties, but without the presence of some level of spatial diversity during any given year, there is still a risk to production. The suggestion from econometric results in Chapter 8 of a potential trade-off between spatial and temporal diversity underline the need for future research regarding the generality and the extent of such a trade-off. If one or the other type of diversity indeed plays a more significant role in productivity, then policy interventions would need to focus on it. However, the ideal combination of the two is not yet clear, nor the methodology to arrive at a solution. We suspect that the balance depends heavily on agro-ecological factors, likely to be region-specific as well as dynamic.

Comparing Measures of Diversity

An "ideal" diversity measure would, at a minimum, be based on molecular differences detectable with certainty at the field/household level. Particularly since no "best" concept or measure capturing all dimensions of diversity is yet evident, the research described in this book has reinforced the importance of different concepts of diversity for different purposes: there are "more" and "less" appropriate concepts for different aspects of analysis. Similarly, there is no "best" measure of diversity, as each measure provides different information and offers a distinct read on diversity and its value or effects. Often different diversity measures were significant under differing circumstances or for different settings, suggesting that there is value in testing varietal richness, inverse dominance and evenness in analyses of spatial diversity. While there are some possibilities for substitution between measures, no obviously redundant measures were identified in the analysis. These findings agree with results from the crop science literature examining relationships between morphology-based and molecular-based diversity measures and which find complementarity, rather than substitutability (Lage et al. 2003). Still, as spatial diversity measures can be correlated, some care is nevertheless required in selecting and including multiple measures in a model specification.

The primary measures used in this volume, as in many other economic studies of diversity, are spatial measures of evenness and dominance. The concurrent analysis of a set of indices and taxonomies at the provincial level in China suggests that latent diversity indices (CODs) tend to exhibit the same trends as morphologically-based (apparent) indices and different trends from those detected using measures of apparent diversity based on named varieties. Precautions on working with named varieties have been repeatedly raised, and we recognize that variety names are rough proxies for differences in the varieties. Whether or not these are actual differences in diversity remains a question, and thus variety-based diversity will always be an imperfect assessment of underlying changes. For apparent diversity, the morphological approach is probably superior to using variety names. The convergence of information from morphological groups and COD measures is promising, but there are still not enough data sets with all three taxonomies to enable the generalization of results. More empirical work in varied systems and environments will provide additional information in refining the use of these and other measures.

Policy Implications

The improved understanding of the influence of policy on diversity reinforces the need to continue assessing the impacts at national and industry levels, as well as at local levels. There can be implications, both foreseen and unforeseen, for diversity from a wide range of policy interventions. Since it is increasingly recognized that diversity has an important role for continued productivity increases, those considering future policy interventions should be aware of potential consequences for diversity outcomes. Through the information obtained from breeders on their perceptions of diversity, their breeding priorities and their constraints, we have a better understanding of the way that the institutional setting influences their priority setting and decision-making process. There are clear linkages between the changes over time in breeders' perceptions about diversity (and thus the supply of genetic diversity) and the associated policy environment. The role of policies affecting markets and infrastructure, as well as the pivotal role of the supply of diversity through plant breeding programs, have also been clearly highlighted.

We can summarize some of the generalizable policy-related findings from the research across both countries as follows:

- Policies focusing on specific traits, such as quality, or specifying quality requirements narrow the genetic base.
- Policies that create more pressure to release varieties more quickly also narrow the genetic base. Additional investments in agricultural research may offset some of these effects.
- More funding lessens constraints, all else equal, to addressing multiple breeding objectives, including diversity.
- Changes that result in more diversity in the production environment (e.g., expansion of irrigated areas or new production technologies) can result in a new set of varieties, but without a guarantee of increased diversity.
- Moving from the household to more aggregate levels of analysis, diversity increases, due largely to increased variation in agro-ecological environments.

If crop diversity is an objective for policy makers, whether it be a social, government or scientifically-motivated objective, ensuring the appropriate policy settings is crucial. Research suggests that there can be a role for policies that specifically address the maintenance or enhancement of diversity. Smale (2005) notes that on-farm conservation of traditional crop varieties occurs insofar as any tradeoffs between diversity and production align with social preferences. These social preferences, the form of the yield-diversity relationship, and farm level production decisions together also determine the social costs and benefits of crop diversity. The principles are no different for predominantly modern and commercial cropping systems. As in the context of traditional varieties, the yield-diversity relationship is a pivotal factor in the degree of priority given to diversity. Where there is a trade-off between yield and crop diversity, the choices for policy makers in countries with current food shortages are particularly acute. However, indications that the yield-diversity relationship, at least at an aggregate level, is perhaps less of a trade off and more of a complementary relationship, suggest that there is much more to gain by ensuring adequate levels of diversity. From the evidence of analyses in this volume, it appears possible to achieve both acceptable yields and diversity outcomes, whatever the production system.

We have also tried to draw out new information on linkages across levels of geographical aggregation through the examination of data sets describing diversity at these different levels. Both practical issues, such as how best to aggregate data from province to national level, as well as broader conceptual issues, needed to be explored. For the China analysis based on morphological characteristics, completely separate analyses at provincial and national levels of the set of morphological traits seemed to make the most sense, while the COD analysis had to be handled somewhat differently. Incomplete data at the shire level prevented a limited generalization to Australia.

At a conceptual level, it is interesting to ask what additional information can be drawn from the increasing pattern of diversity observed between subunits at successively higher levels of aggregation over space (e.g., household to village, shire to state, province to country) and what this pattern implies in terms of linkages among the various decision makers and institutions involved in determining diversity outcomes. An ideal data set would be carefully constructed to overlay variety information at the household level in increasing levels of aggregation. Due to the lack of more complete information linking village and county levels and county and provincial levels, we were unable to examine adequately and fully the linkages up from the household level that would shed more light on the smallest unit for which it is appropriate to focus policy attention. We suspect, however, that linkages are likely to be country and context specific and thus may be difficult to generalize.

Certainly there will be different implications for systems characterized predominantly by modern varieties, where crop diversity primarily counteracts production risks at more aggregated levels, as opposed to systems with traditional varieties, where conservation considerations arise in addition to both production and consumption concerns at household and community levels. We must also consider that, while diversity outcomes generally increase as the level of aggregation increases, the certainty with which we can attribute changes in diversity outcomes to behavior resulting from specific policies or institutional structures decreases, as we move away from on farm and other micro-level models. These micro-level models of variety choice and diffusion will continue to be important given that genes are embodied in crop varieties and farmers "choose" genes indirectly through their selection of crop varieties. It is this interaction of farmers' decisions with the larger physical and socioeconomic environment that drives the economic analysis of crop technologies, including agricultural biodiversity and biotechnology.

Whether the production system is primarily traditional or modern also carries implications for the scale and level of intervention points to support crop diversity. Traditional systems are characterized by a high degree of local adaptation in both agronomic and consumption traits, so that interventions at an appropriately local level likely need to be considered. Market participation on the part of household farms takes place, but participation is uneven and can be irregular even among participating households. Policy intervention through markets is only as effective as the level of market participation and the operating efficiency of the markets. Issues of local adaptation are also relevant to modern systems, but in general, local preferences and characteristics do not play as large a role as in traditional systems. Interventions through markets, as well as at more aggregate levels in the seed system, may thus be an effective way to influence diversity.

Further understanding will come with additional research. In particular, as it is possible to draw on more detailed data and develop models more fully to incorporate linkages across increasingly disaggregated levels of analysis, hypothesis tests may be specified for more targeted policy issues at the local, regional and national levels. The methodologies proposed in this volume may also prove useful for analysis of the spatial distributions of varieties carrying certain types of genetic resistance to disease or endowed with transgenes. Further investigation of the relationship and potential tradeoffs between spatial diversity and temporal diversity is also likely to be valuable, as well as improved measures of temporal diversity to be used in dynamic analyses of diversity.

REFERENCES

Lage, J., M.L. Warburton, J. Crossa, B. Skovmand, and S.B. Anderson. 2003. Assessment of genetic diversity in synthetic hexaploid wheats and their *Triticum dicoccum* and *Aegilops tauschii* parents using AFLPs and agronomic traits. *Euphytica* 134: 305-317.

Smale, M. 2005. Chapter II: Economics literature about crop biodiversity: Findings, methods and limitations. In *The State of Economics Research: Sustainable Management of Crop and Livestock Biodiversity*. Report prepared for the CGIAR System-wide Genetics Resources Program.